Sorghum Medicinal Food
- Medicinal and Industrial Perspective -

Author

Dr. U.D. Chavan obtained his B. Sc. and M. Sc. (Agri. in Biochemistry) degrees from Mahatma Phule Krishi Vidyapeeth Rahuri in 1985 and 1987 respectively. He received his Ph.D. degree in Food Science from Memorial University of Newfoundland St. John's Canada in 1999. He has done International training on "Global Nutrition 2002" at Uppsala University Uppsala, Sweden in 2002. He also attended follow-up International workshop on "Global Nutrition 2002" at Hanoi, Vietnam in 2002. Dr. Chavan visited Denmark, Finland, Ireland, France, Switzerland, Poland, Spain, Vietnam, Thailand, England, and U.S.A. under "Global Nutrition 2002" Programme sponsored by Swedish International Development Agency (SIDA). During this programme he worked on human nutritional diet and disorders as well as on genetically modified organisms (GMO).

Dr. Chavan worked as Senior Research Assistant in the Department of Biochemistry and Food Science and Technology at Mahatma Phule Krishi Vidyapeeth Rahuri from 1988 to 2000. During his Ph. D., he worked as Technician/Research Associate at Atlantic Cool Climate Crop Research Center and Agriculture and Agri-Food Canada. He received D.Sc. degree in 2006 from USA. He has guided 25 students for M.Sc. (Agri.) in Biochemistry and Food Science and Technology. From 2000 to 2004 he worked as Assistant Professor of Biochemistry at Mahatma Phule Krishi Vidyapeeth Rahuri.

Dr. Chavan received Korgavkar Trust Fellowship, under-graduate and post-graduate merit scholarship as well as Senior Research Fellowship from ICAR, New Delhi and United States Department of Agriculture U.S.A. He received University Graduate Fellowship from Memorial University of Newfoundland St. John's Canada for his Ph. D. programme. Dr. Chavan also received international Scholar Award and Excellentiam Award for his Ph.D. research work from Memorial University of Newfoundland St. John's Canada. He has received a "Certificate of Appreciation" from the U.S. Department of Agriculture in 1992 for his work on processing of groundnut under the guidance of Dr. S.S. Kadam (Principle Investigator). He has written 135 research papers and 140 popular articles. He has authored 19 books in Marathi, 23 books in English and eight book chapters in English. Dr. Chavan has visited 23 countries in the world for research and teaching programmes. He has been awarded **"Literary Award"** for his best book on **"Growth Regular"** on agriculture during **1997** by Government of Maharashtra. Dr. Chavan was selected as a best group leader and best presentation for "Global Nutrition 2002" by SIDA. He has been awarded "Life Time Achievement Award 2004" for his outstanding contribution in post-harvest technology of fruits and vegetables and allied fields by United Writers' Association of India. He has been elected as a Fellow of United Writers' Association of India in 2004. Dr. Chavan received pride of nation and Maharashtra Gunigan Ratna Award for 2006. He also received Jewels of India Award for 2006 for his contribution in the field of Food Science and Technology. He has been awarded **"Literary Awards"** for his best book on **"Sorghum Grain Processing"** on agriculture during **2009-2010** and **"Pulses cultivation to processing and value addition"** for **2012** by Government of Maharashtra. He is recipient of "Krishi Goorav Award-2011" from Bhartiya Krishak Samaj, Maharashtra State, Nasik. He also contributed in the development of crop varieties in wheat NIAW-917 and in Sorghum Phule Anuradha (RSV 423), Phule Chitra (RSV 1546), Phule Suchitra (RSV 1098), Phule Vasudha (RSV 423), Phule Revati (RSV 1006), SPV-1626, Phule Panchami (RSPOV-3) for Sorghum pops, Phule Godhan (SPV-2057) for forage single cut, CSH-50 (Sweet Sorghum), Phule Madhur (RSSGV – 46) for sorghum *hurda* and Phule Rohini (RPASV-3) for sorghum *papad*. Now he is working as a Head of the Department of Food Science and Technology, as well as a Senior Cereal Food Technologist in Sorghum Improvement Project and Foreign Student Advisor at Mahatma Phule Krishi Vidyapeeth, Rahuri, Dist, Ahmednagar, Maharashtra, India.

Sorghum Medicinal Food

- Medicinal and Industrial Perspective -

U.D. Chavan

Head

Department of Food Science and Technology

Mahatma Phule Krishi Vidyapeeth, Rahuri

SCIENTIFIC PUBLISHERS

Published by:

SCIENTIFIC PUBLISHERS (INDIA)
5A, New Pali Road, P.O. Box 91
Jodhpur 342 001, India
E-mail: info@scientificpub.com
Website: www.scientificpub.com

ISBN: 978-93-89412-56-7
eISBN: 978-93-89412-57-4

Printed in India

Content

Foreword

The relationship between diet, moderate physical activity and health is of much current interest. Although initial evidence for this relationship was mainly epidemiological in nature, recent advances and developments in the area of nutraceuticals and functional foods have led to substantiation of claims by preclinical and clinical studies. Furthermore, availability of sophisticated analytical tools has led to the identification and appreciation of non-nutrient bioactives in the diet. Another motivating factor has been mounting health care costs that have emphasized a need for a preventive rather than a treatment approach by consumers, health professionals, researchers and perhaps some government departments in certain countries. The emerging findings in nutrigenomics and proteomics and the advent of nanotechnology for better delivery of bioactives have opened the way for individualized dietary regimes in the hopefully not too distant future. Over the past decade or two there has been a surge in the publication of research findings and developments in the field, although often in a non-targeted and fragmentary manner. Therefore, to fill the existing gap, the launch of a dedicated book for **"Sorghum Medicinal Food** (Medicinal and Industrial Perspectives)" was deemed necessary. This idea was first come in the mind of author and Astral Publication committed to publish this type of book. In this book medicinal properties of sorghum and value added products from sorghum are given in details. I think author has chosen the wright subject of the book and I hope this book will serve the purpose of students, teachers, readers, consumers and professional industrial processors.

J. K. Chavan

Ex-Head FST

MPKV, Rahuri

Preface

Sorghum contains various phytochemicals (including phenolic compounds, plant sterols and policosanols) that are secondary plant metabolites or integral cellular components. Phenols help in the natural defense of plants against pests and diseases, while the plant sterols and policosanols are mostly components of wax and plant oils. The phytochemicals have gained increased interest due to their antioxidant activity, cholesterol lowering properties and other potential health benefits. The phenols in sorghums fall under two major categories; phenolic acids and flavonoids. The phenolic acids are benzoic or cinnamic acid derivatives, whereas the flavonoids include tannins and anthocyanins as the most important constituents isolated from sorghum to date. Sorghum phytosterols are similar in composition to those from corn and contain mostly free sterols or stanols and their fatty acid ferulate esters. The sterols and stanols are structurally similar, except for the presence of a double bond at position 5 in sterols, which is lacking in stanols. The policosanols (fatty alcohols) exist mostly as free or esterified forms with C_{24}–C_{34} atoms, and the general formula $CH_3–(CH_2)_n – CH_2OH$. In sorghum-free forms of the C_{28} (octacosanol) and C_{30} (triacontanol) are the most abundant.

Despite the high levels and diversity of phytochemicals in sorghum, research on this crop as a source of valuable health promoting compounds lags behind other commodities (e.g., fruits and vegetables). As a result, utilization of sorghum fractions in foods to improve nutrition is very limited. Sorghum has a big potential, given its agronomic properties, as well as the emerging evidence on the biological effects of the phytochemicals present in the grain. The purpose of this book is to demonstrate that sorghums with special characteristics exist that have potential as significant sources of condensed tannins, anthocyanins and other phytochemicals with properties that complement the phytochemicals present in fruits and vegetables.

U.D. Chavan

Acknowledgement

Author is thankful to the Hon. Dr. S. S. Kadam, Hon. Ex. Vice-Chancellor, Marathwada Agricultural University, Parbhani and Ex. Head Department of Food Science and Technology, Mahatma Phule Krishi Vidyapeeth, Rahuri for his critical and fruitful suggestions while the preparation of **"Sorghum Medicinal Food** (Medicinal and Industrial Perspectives)" book. I am thankful to Dr. Fereidoon, Shahidi, Dr. J. K. Chavan, Dr. Amarowicz Ryszard, Dr. J. V. Patil, Dr. V. D. Pawar, Dr. U. Y. Bhoite, Dr. K. D. Chavan, Dr. C. A. Nimbalkar, Mr. Ajay Jadhav, Mr. Kashinath Bagale, Mr. Suresh Shedage for helping during this manuscript preparation, as well as those scientists who worked on phenolic antioxidants and health benefits identification, utilization and usefulness for human beings and their contribution is acknowledged in this book as well their sources are cited in the reference section. Authors are also thankful to all scientists and friends those helped directly or indirectly while preparation of this manuscript. We would like to express over profound thanks to our family members for their constant support, patience and devotion, which inspired me to complete this manuscript.

U.D. Chavan

Chapter – 1

Introduction

Sorghum (*Sorghum bicolour* L. Moench) is a genus of numerous species of grasses, one of which is raised for grain and many of which are used as fodder plants either cultivated or as part of pasture. The plants are cultivated in warmer climates worldwide. Species are native to tropical and subtropical regions of all continents in addition to the southwest Pacific and Australasia. Sorghum is in the subfamily Panicodeae and the tribe Andropogoneae (the tribe of big bluestem and sugar cane). Sorghum is divided in 29 species and two hybrids as follows:

SORGHUM SPECIES

The known species of the sorghum are listed here for information to the students, teachers and researchers for future development in the field of sorghum research. *Sorghum almum; Sorghum amplum; Sorghum angustum; Sorghum arundinaceum; Sorghum bicolour* cultivated sorghum, often individually called sorghum. Also known as durra, jowari, or milo; *Sorghum bicolour subsp. drummondii Sudan grass; Sorghum brachypodum; Sorghum bulbosum; Sorghum burmahicum; Sorghum ecarinatum; Sorghum exstans; Sorghum grande; Sorghum halepense* Johnson grass; *Sorghum interjectum; Sorghum intrans; Sorghum laxiflorum; Sorghum leiocladum; Sorghum macrospermum; Sorghum matarankense; Sorghum nitidum; Sorghum plumosum; Sorghum propinquum; Sorghum purpureosericeum; Sorghum stipoideum; Sorghum timorense; Sorghum trichocladum; Sorghum versicolour; Sorghum verticiliflorum; Sorghum vulgare* var. *technicum* broomcorn and two hybrids *Sorghum x almum*; *Sorghum x drummondii*.

One species, *Sorghum bicolor*, is an important world crop, used for food (as grain and in sorghum syrup or sorghum molasses), fodder, the production of alcoholic beverages and biofuels. Most varieties are drought and heat-tolerant, and are especially important in arid regions, where the grain is one of the staples for poor and rural people. These varieties form important components of pastures in many tropical regions. *Sorghum bicolor* is an important food crop in Africa, Central America and South Asia and is the fifth most important Cereal crop grown in the world after rice, wheat, maize and barley.

Sorghum is one of the major cereal crops among the all millets grown mostly in dryland. It is grown as rainfed crop in both season *Kharif* (mansoon) and *Rabi* (winter). It is mainly grown in the Deccan plateau, Central and Western India apart from a few patches in northern India. Almost entire grain produce is used for human consumption in India. It is nutritionally superior to other fine cereals such as rice and wheat and hence it is known as nutritious cereal. Nutritionally sorghum grain contains 4.4 to 21.1% protein, 2.1 to 7.6% fat, 1.0 to 3.4% crude fiber, 57.0 to 80.6% total carbohydrates, 55.6 to 75.2% starch and 1.3 to 3.5% total minerals (ash). Sorghum also provides 350 Kcal energy, calcium, phosphorus, potassium, carotene and thiamin as well as antioxidants through phenolics and various types of tannins. It is mainly consumed as *Bhkari* (*roti*) i.e. un-leavened pancake in various states of India. Apart from the traditional products like *bhakari, bhatwadi, papad, popped grains, kurdai, high fibre cookies, biscuits, flakes, thalipeeth, upama, rawa idali, dosa, uttappa, chiwada, chakali, papadi, ambil, shankarpale, cookies and cakes* are also prepared from sorghum and consumed as snack food items. Sorghum grains are not only good source of nutrients but it also contains special constituents such as phyto-chemicals, dietary fbire as well as resistant starch, which are more essential to the human nutrition. However, the grain sorghum consumption has remained restricted to the weaker sections of the society due to poor nutritional quality of grain and inferior quality of the products such as Bhakari as well as very low price of this food item as compared to the other cereal grains such as wheat, rice and other millets.

Sorghum is of African origin. A large variety of wild and cultivated sorghum are grown in the tropics and subtropics of the world. In India, sorghum constitutes an important article of food, after rice and wheat. The sorghum grain is small and rounded, varying in colour from off-white to white to varying shades of red, yellow or brown. The grain size varies, the weight ranging from 7.0 to 61 g/1000 grains, with most sorghums weighing 20 – 30 g/1000 grains. The chemical composition of grain sorghum is similar to that of maize. Generally, sorghum has more protein than maize, a lower fat content and about the same amount and proportions of carbohydrate components. The proximate analysis of Indian sorghum grain indicates moisture, 11.9; protein 10.4; fat 1.9; fiber 1.6; carbohydrates 72.6 and minerals 1.6; minerals present in the grain are calcium, magnesium, potassium and iron.

In comparison with maize, sorghum grain contains approximately the same quantities of riboflavin and pyridoxine but more pantothenic acid, nicotinic acid and biotin. Nicotinic acid occurs in the grain in available form.

Starch is the major carbohydrate of the grain. The other carbohydrates present are simple sugars, cellulose and hemicelluloses. The amylose content of starch varies from 2 1 to 28%. Starch from waxy varieties contains little amylose. Both waxy and regular starches contain free sugars upto 1.2%. Sucrose being the major constituent (0.85%) followed by glucose (0.09%), fructose (0.09%), maltose and stachyose. Sorghum grain contains no detectable amount of glucoside, but on germination within 3 days old seedling gives 3.5% Dhurin which leads to the poisoning of animals consuming such sorghum seedlings.

The percentage of different protein fractions to the total protein of sorghum grown in India is albumin 5; globulin 6.3; prolamin 46.4 and glutelin 30.4. Prolamin and glutelin are principally present in the endosperm. Amino acid analysis of various protein fractions shows that there is better distribution of all essential amino acids in globulins than in prolamins. Sorghum protein is superior to wheat protein in biological value and digestibility. A vegetarian diet based on some varieties of sorghum is somewhat better than a rice-based diet. Sorghum lipids mostly consist of triglycerides, which are rich in the unsaturated fatty acids, oleic and linoleic, their percentage being 33 and 47 respectively.

The nutritional quality of grain sorghum can be improved by malting, fermentation and by mixing of flour from other cereal grains or pulses. Malting or germination of sorghum grain have been reported to improve the nutritional quality in terms of increased soluble proteins, free amino acids, reducing sugars, digestibility of proteins and starch. Such type of grain processing treatments increases dietary fibre content in the final processed products. The higher proportion of dietary fibre of grain sorghum confer many advantages such as nutritional, digestive and physiological benefits such as hypocholesterolemic and hypoglycemic effects, incidence of colon cancer, constipation and gastro-intestinal complications etc. A large number of diabetic patients are regularly consuming sorghum bhakari with fenugreek or bitter gourd *subji* through every day meals. However, the bhakari prepared from malted sorghum flour has been found to contain higher proportion of dietary fibre with superior nutritive and digestive values with additional physiological benefits. Natural sorghum roti consumption has been restricted to only weaker section of the society due to its coarse nature and palatability. This type of problems can be overcome by using malting and fermentation technology for sorghum.

Sorghum is undoubtedly the staple cereal of Botswana and is grown widely in the country. Statistical surveys indicated that 96% population of southern and Southeastern districts of Botswana mainly consuming sorghum as a major food. Several varieties of sorghum and millet, both indigenous and exotic are grown in the country. All these varieties have proved to be of good food value and may be used for different industrial applications. The development of these varieties and their subsequent use has been steered by a number of institutions in the country and abroad, notably the ministry of Agriculture, SADC/ICRISAT, the Botswana Technology Center (and its international collaborators-CIRAD, NRI and IICT) and others. An identified constraint that continues to higher the national objective of food security is postproduction losses. These losses are believed to be high in magnitude although there is no loss assessment studies carried out in the country. These losses are caused by a number of agents, mainly insects and postproduction practices.

In many parts of the world sorghum has traditionally been used in several food products and various food items, like porridge, unleavened bread, cookies, cakes, couscous and malted beverages are made from this versatile grain. The traditional food preparation of sorghum is quite varied according to the region. Boiled sorghum is one of the simplest uses and small, corneous grains are normally desired for this type of food product. The whole grain may be ground in to flour or decorticated before grinding to produce a fine particle product or flour, which is then used in various traditional foods. Sorghum is a good source of lactobacilli, which is used in souring of foods used in traditional fermented foods and drinks. Nigerians produce "*Ogi*" a traditional fermented thin porridge. Fermented breads called "*masa*" and unleavened bread called "*waina*" are also produced in Nigeria. "*Tuwo*" is a stiff porridge produced in Nigeria from adding boiling water to sorghum flour and continuously stirring. Couscous can be made from sorghum. The grain is also used in making starch, dextrose, syrup, gur, vodka, malted flour for weaning foods, beers, and other alcoholic beverages.

Sorghum playing an important role in animal feed production in various countries the USA, Mexico, South America and Australia. The nutritional value of good quality sorghum has an average feeding value of 96 to 98% of corn. Sorghum can be processed to further improve its food as well as feed value and techniques such as grinding, crushing, steaming, steam flaking, popping and extruding have all been used to enhance the grain for food and feeding purpose. The products are then fed to beef or dairy cattle; egg laying hens, poultry, and pigs are also used in pet foods.

Industrial uses for sorghum include wallboard and biodegradable packaging materials. In addition to the wallboard grain sorghum can be processed interchangeably with corn for the production of ethanol-many ethanol producing factories within the United States use grain sorghum as their principal grain due to their favourable geographic location. A very valuable co-product created from the processing of grain sorghum for ethanol is dry distillers' grain. Grain sorghum having high protein animal feed supplement generates additional revenues for the ethanol processing plant.

BENEFITS OF SORGHUM CONSUMPTION

1. Sorghum food provides non-glutinous flours which is useful for avoiding damage to the lining of the intestine and make easy for nutrients to be absorbed
2. It avoid celiac disease problems
3. It also avoid dermatitis herpetiforms disease
4. Sorghum is non-acid forming food
5. Sorghum foods are a soothing and easy to digest
6. Sorghum foods are least allergenic and mostly digestible
7. It is good source of dietary fibers
8. It also good source of vitamins and minerals

9. It also provides antioxidants for controlling cancers
10. Malted sorghum flour can be used for anti-constipation drinks
11. It provides sufficient amount of dietary fiber in human diet
12. Selected sorghum genotypes can be provide enough amount of lysine to meet human requirement
13. Sorghum diet is very good to the people suffering from the Jaundice
14. Sorghum has very low glycimic index so it helps to human health
15. Sorghum helps in loose and control of body weight
16. It increases body's sensitivity to insulin
17. It also helps in diabetes control
18. It reduces risk of heart diseases
19. It also reduces blood cholesterol levels
20. It also reduces hunger and keeps fuller for longer time
21. It also helps to manage polycystic ovary syndrome in women

ANTI NUTRITIONAL FACTORS AND MYCOTOXINS IN SORGHUM

Sorghum contains polyphenols, which are generally associated with grain pigmentation. Polyphenols commonly referred to as tannins, interfere with the bioavailability of nutrients. It has been reported that sorghum grains contain substantial quantities of condensed tannins and trace amounts of hyrdrolyzale tannins. Brown sorghums with high tannins are known to be resistant to bird depredation and have reduced preharvest germination and grain moulding.

Sorghum grain is known for its hardness compared to other food grains. The hardener of the grain is due to higher content of protein prolamin. Prolamin content varies from 36.0 to 51.0%. The antinutritional factors present in sorghum grain are polyphenols, and phytic acid. Polyphenols are the secondary metabolites produced and the presence of polyphenols/tannins in diet result in poor digestibility of proteins as they bind the proteins present in grain and make them unavailable for the intestinal absorption. Mostly polyphenols present in colored sorghums. In white sorghum, flavan 3-ols or flavan 4-ols (monomers of polyphenols) are present in very low quantity. Apart from flavan 4-ols, phytic acid is also present in sorghum in the form of 6-inositol phosphate. Phytic acid usually forms insoluble compounds with minerals like calcium, iron, magnesium and zinc thus making them unavailable.

Presently the new research investigations on polyphenols and phytic acid consider these compounds as health factors and consumption of these factors increase immunity in animal and human systems against several diseases. Due to grain mould the kharif grain gets severely affected by *Fusarium* and *Aspergillus* fungi, which produce harmful toxins. These mycotoxins namely aflatoxins and fumonisins cause deleterious effects to human and animal health. Hence the grain that is used for food preparation and consumption should be free of these toxins. However these toxins have a safety limit beyond which they cause ill health. The safety limit for aflatoxins is 20 ppb and for fumonisins it is 200 ppb as per the CODEX committee. Aflatoxin contamination was relatively less compared to Fumonisin contamination.

COMPARISON WITH OTHER CEREALS

Grain sorghum is mostly used for human food as well as in small quantity for animal feeds. The utilization of sorghum grains in the world is shown in **Table 1.1**. From this table it is conformed that most of the sorghum grains are used as food in the Asia and Africa. Developed countries are using very small quantity of sorghum grains as a food and remaining mostly using for animal feeds.

Table 1.1: Classification of sorghum utilization (average million tones).

Region	Food	Feed	Other uses	Total
Africa	8.0	0.4	2.3	10.7
Asia	15.1	6.3	2.1	23.5
Central America	0.3	8.4	0.2	8.9
South America	-	4.6	0.3	4.9
North America	-	12.6	0.1	12.7
Europe	-	1.4	-	1.4
USSR	-	2.3	0.3	2.6
Oceania	-	0.4	-	0.4
World	23.4	36.4	5.3	65.1
Developing countries	23.2	15.6	4.8	43.6
Developed countries	0.2	20.8	0.5	21.5

Sorghum grains are rich in fiber and minerals apart from having a sufficient quantity of carbohydrates, proteins, and fat. The comparison of sorghum grain with other cereals is given in following **Tables 1.2, 1.3, 1.4, 1.5, 1.6 and 1.7.**

Table 1.2: Average composition of cereal grains.

Grains	Moisture (%)	Crude Protein (%)	True protein (%)	Oil/ Ether extract (%)	Crude fiber (%)	CHO (NFE) (%)	Ash (%)
Barley	15.0	9.0	8.5	1.5	4.5	67.4	2.6
Sorghum	11.0	9.6	8.6	3.8	1.9	71.3	2.4
Maize	13.0	9.9	9.4	4.4	2.2	69.2	1.3
Millet	13.0	10.5	9.9	3.9	8.1	60.7	3.8
Oats	13.0	10.4	9.5	4.8	10.3	58.4	3.1
Rice	11.4	8.3	7.2	1.8	8.8	64.7	5.0
Rice polished	13.0	6.7	6.4	0.4	1.5	77.6	0.8
Rye	13.0	11.6	10.7	1.7	1.9	69.8	2.0
Wheat	13.0	12.2	11.0	1.9	1.9	69.3	1.7

Table 1.3: Amino acid composition of sorghum, wheat and rice grains (g/16 g Nitrogen).

Amino acid	Sorghum	Wheat	Rice
Lysine	2.60	2.72	3.68
Histidine	2.06	2.08	2.08
Arginine	4.20	4.64	7.68
Aspartic acid	7.21	3.40	4.85
Threonine	3.00	2.88	3.68
Serine	3.66	4.30	4.56
Glutamic acid	20.30	32.50	11.69
Proline	6.25	11.60	5.27
Glycine	3.45	3.20	6.41
Alanine	8.64	2.00	3.23
Cystine	0.97	2.24	1.44
Valine	4.39	4.48	6.08
Methionine	1.39	1.44	2.40
Isoleucine	3.83	3.52	4.80
Leucine	12.27	6.56	8.00
Tyrosine	3.25	2.88	4.64
Phenylalanine	4.55	4.48	4.48
Trptophan	1.12	1.12	1.28
Total	93.14	96.04	86.25
Energy (Kcal)	349	341	345

Table 1.4: Mineral composition of sorghum, wheat and rice flour (mg/100 g).

Mineral	Sorghum	Wheat	Rice
Sodium	21.00	19.3	2.2
Potassium	537.00	315.0	7.4
Calcium	25.00	48.0	10.0
Phosphorus	526.00	183.0	160.0
Magnesium	212.00	132.0	90.0
Iron	8.48	4.90	0.70
Zinc	3.91	2.20	1.40
Copper	0.86	0.51	0.14
Manganese	3.50	2.29	0.59

Table 1.5: Vitamins content in sorghum, wheat and rice grains.

Vitamin	Sorghum	Wheat	Rice
β-Carotene (μg/100g)	47.00	64.00	-
Thiamine (mg/100g)	0.37	0.45	0.06
Riboflavin (mg/100g)	0.13	0.17	0.06
Niacin (mg/100g)	3.10	4.30	1.90
Total B6 (mg/100g)	0.21	0.57	-
Folic acid (μg/100g)	20.00	35.80	8.00

Table 1.6: Oxalic acid, phytic phosphorus and dietary fiber in sorghum, wheat and rice grains.

Constituent	Sorghum	Wheat	Rice
Oxalic acid (mg/100g)	10.00	8.00	3.00
Phytic Phosphorus (mg/100g)	172.00	238.00	83.00
Phytic P as percent of total P	77.00	80.00	52.00
Total dietary fiber (%)	12.69	11.40	-

Table 1.7: Synergistic effects of legume (green gram) on PER of sorghum.

Source of protein	PER
Casein	2.87
100% Sorghum	0.93
90% Sorghum + 10% green gram	1.11
80% Sorghum + 20% green gram	1.36
70% Sorghum + 30% green gram	1.75
50% Sorghum + 50% green gram	1.56

REASONS WHY WE HAVE TO EAT GRAIN SORGHUM

1. Grain sorghum tastes good Sorghum's neutral flavour blends well with the flavours of additional ingredients.
2. Grain sorghum is available as whole grain whether popped as a snack, milled into flour or cooked whole as a cereal or pilaf.
3. Grain sorghum is high in potassium (350 mg/100 g) and low in sodium (6 mg/100 g) therefore promoting healthy blood pressure.
4. Grain sorghum is gluten-free enabling those with celiac disease to consume a healthy whole grain product.
5. Grain sorghum is rich in health promoting phytochemicals: phenolic acids, sterols, policosanols, and anthocyanins.
6. Grain sorghum and sorghum flour are rich in minerals: magnesium, copper, manganese, iron and zinc.
7. Grain sorghum and sorghum flour are rich in vitamins: thiamin, riboflavin, niacin, vitamin B-6 and pantothenate.
8. Grain sorghum is rich in macronutrients: containing 11% protein, 75% carbohydrate and 3% total fat.
9. Grain sorghum grown in the *Rabi* season is free/lower level of condensed tannins that may interfere with mineral absorption.
10. Grain sorghum and sorghum flour are excellent sources of fiber ranging from 6.3-11.1g and 6.6g per 100g respectively.

GLUTEN-FREE WHOLE GRAIN SORGHUM AND EXCELLENT SOURCE OF DIETARY FIBER

Whole grain gluten-free sorghum flour is an excellent source of dietary fiber with 6.6g/100g flour (USDA National nutrient database for standard reference, Release 24 (2011), item No. 20648). Concentrated sources of fiber include whole grains, dried peas and beans, vegetables, nuts and fruits. Fiber is intrinsic and intact only in plants. Dietary fibers are non-digestible forms of carbohydrates and lignin. Dietary fiber can be classified as fermentable or non-fermentable both forms are needed for good health. Along with helping provide satiety, a moderate amount of evidence suggests that dietary fibers from whole food sources protect against cardiovascular disease, obesity and type 2 diabetes and is essential for optimal digestive health (http://www.cnpp.usda.gov/publications/ dietaryguidelines/2010/dgac/report/d-5-carbohy drates.pdf). In particular, individuals with celiac disease may not consume enough dietary fiber and need gluten-free whole grains such as sorghum in their diets.

Based upon an extensive review of the evidence, the FDA recently approved a health claim for whole grains that permits manufactures to state on foods: Diets rich in whole grain foods and other plant foods and low in total fat, saturated fat and cholesterol may reduce the risk of heart diseases and some cancers. In order to use the claim, the serving size must contain a specified amount of dietary fiber per serving. Four serving sizes are specified ranging from a 35 grams serving with 1.7 grams of fiber to a 55 gram serving with 3.0 grams of dietary fibers. Thus whole grain sorghum is an excellent source of dietary fiber. The nutritional composition of grain sorghum are compared with similar other cereal grains (**Table 1.8**).

Table 1.8: Nutritional composition of sorghum compare with other cereals (100g).

Nutrient	Sorghum	Wheat	Corn	Rice	Cassava	Amaranth	Quinoa
Energy, Kcal	339	342	365	360	160	371	368
Protein, g	11.30	11.31	9.42	6.61	1.36	13.56	14.12
Total fat, g	3.30	1.71	4.74	0.58	0.28	7.0	6.1
Carbohydrate, g	74.60	75.90	74.26	79.34	38.06	65.25	64.16
Fiber, g	2.7	12.2	7.3	n/a	1.8	6.7	7.0
Calcium, mg	28.0	32.0	7.0	9.0	16.0	159.0	47.0
Iron, mg	4.4	4.56	2.71	0.81	0.27	7.6	4.57
Magnesium, mg	190	93	127	35	21	248	197
Phosphorus, mg	287	355	210	108	27	557	457
Potassium, mg	350	432	287	86	271	508	563
Sodium, mg	6	2	35	2	14	4	5
Zinc, mg	1.54	3.33	2.21	1.16	0.34	2.9	3.1
Copper, mg	1.080	0.363	0.314	0.110	0.100	0.52	0.29
Manganese	1.630	3.821	0.485	1.100	0.384	3.33	2.03
Iodine, μg	n/a	n/a	n/a	n/a	n/a	n/a	n/a
Selenium, mcg	Trace	8.3	15.5	n/a	0.7	18.7	8.5
Vitamin C, mg	2	0.0	0.0	0.0	20.6	4	n/a

Nutrient	Sorghum	Wheat	Corn	Rice	Cassava	Amaranth	Quinoa
Thiamine, mg	0.237	0.387	0.385	0.070	0.087	0.116	0.360
Riboflavin, mg	0.142	0.108	0.201	0.048	0.048	0.200	0.318
Niacin, mg	2.927	4.381	3.627	1.600	0.854	0.923	1.520
Pantothenate, mg	1.25	0.954	0.424	1.342	0.107	1.457	0.772
Vitamin B, mg	0.59	0.368	0.622	0.145	0.088	0.591	0.487
Foliate total, mcg	0.02	38.0	19.0	9.0	27.0	82	184
Vitamin B-12, mcg	0.0	0.0	0.0	0.0	0.0	0.0	0.0
Vitamin A, IU	16	9	214	n/a	13	2	14
Vitamin D, µg	n/a	0.0	0.0	0.0	0.0	0.0	0.0
Vitamin E, mg-ATE	1.2	1.01	0.49	n/a	0.19	1.19	2.44

Source: WHO/FAO/UNU. 2007.

MACRONUTRIENTS

Sorghum is an excellent source of energy, containing about 75 percent complex carbohydrate. Complex carbohydrates (fibers and starches) are usually digested slowly and therefore provide satiety and delayed hunger. Sorghum flour is often whole grain and thus aligns well with the 2010 Dietary Guidelines to increase dietary whole grains while keeping the suggested total dietary carbohydrate intake the same (Report of the Dietary Guidelines Advisory Committee on the Dietary Guidelines for Americans, 2010 USDA). A Nutrient comparison of the commodity grains sorghum, wheat, corn, and rice shows that with about 3.4 percent fat content, grain sorghum contains more fat than wheat and rice, but slightly less than corn. Sorghum and wheat grain contain similar amounts of protein while both contain more protein than rice and corn (USDA (2011). National Nutrient Database for Standard Reference, Release 24). Sorghum is naturally gluten-free.

MICRONUTRIENTS

A comparison of 100 g of commodity sorghum to the World Health Organization (WHO) Recommended Nutrient Intakes (RNI) (FAO/WHO, 1998; FAO/WHO/ UNU, 2001; WHO/FAO/UNU. 2007; Food and Nutrition Board. 2001) for children ages 1-3 years shows the following micronutrients are met: Magnesium = 366 percent RNI; Iron based on 10 percent bioavailability = 73 percent RNI; Zinc based on moderate bioavailability = 38 percent RNI; Thiamin = 47 percent RNI; Riboflavin = 28 percent RNI; Niacin = 49 percent RNI; Pantothenate = 63 percent RNI; Vitamin B-6 = 118 percent RNI. WHO does not have an established

RNI for copper and manganese, thus using the United States Recommended Dietary Allowance (RDA)6 criteria, 100g sorghum meets the RDA for children ages 4-8 years as follows: Copper = 245 percent RDA; Manganese = 92 percent RDA. Iron and zinc are two of the four micronutrients (iron, zinc, iodine, vitamin A identified (Committee on Micronutrient Deficiencies, Prevention of Micronutrient Deficiencies, 1998) as at risk in populations of developing countries. Sorghum is a good to excellent source of iron and zinc. Sorghum is rich in B complex vitamins that play a major role in energy metabolism. Sorghum's high-energy content and ready supply of B-complex vitamins are a perfect combination for energy utilization.

PHYTOCHEMICALS

Depending on the variety, sorghum provides good to excellent sources of such phytochemicals as phenolic acids, anthocyanins, phytosterols and policosanols. These compounds are familiar to the public as a result of health claims around sterols and stanols (heart health) and the publicity attributed to the anti-oxidant properties of anthocyanins (pigmented berries-blueberries, strawberries, etc.). Awika and Rooney 8 provide an excellent review of the potential health benefits of phytochemicals in sorghum. One category of phytochemicals, condensed tannins regardless of grain colour is not found in U.S. sorghum varieties and most sorghum produced elsewhere (Awika and Rooney, 2004).

Sugar-slashing nutrients and the drugs that imitate them.

Nutrient	Mechanism of action	Beneficial results	Drug with similar mechanism	Potential drug side effects
Sorghum	Reduces starch breakdown to sugar in intestine by inhibiting amylase and alpha-glucosidase	Prevents glucose from reaching bloodstream	Acarbose	Gastro-intestinal symptoms (pain, gas, bloating)
	Activate PPAR-gamma	Increases insulin sensitivity, does not boost insulin secretion	Thiazolidi-nediones	Fluid retention (edema), cardiovascular risk (rosiglitazone), bladder cancer (pioglitazone)
	Reduces new sugar manufacture in liver (inhibits hepatic gluconeogenesis)	Reduces contribution of newly made sugar to fasting blood glucose	Metformin	Metformin and the potential risk of lactic acidosis are controversial but evolving. Heart & kidney failure.

REFERENCES

Au, P.M. and Fields, M.L. (1981). Nutritive quality of fermented sorghum. J. Food Sci. 46: 652-654.

Awika, J.M. and Rooney, LW. (2004). Sorghum phytochemicals and their potential impact on human health. Phytochemistry 65: 1199-1221.

Badi, S.M. and Hoseney, R.C. (1976). Use of sorghum and pearl millet flours in cookies. Cereal Chem. 53: 733-738.

Bankar, J.R., Chavan, J.K. and Kadam, S.S. (1986). Effect of grain conditioning on physico-chemical properties of sorghum bhakari. J. Maharashtra Agric. Univ., 11: 221-223.

Bender, A.E. (1973). Composition of infant foods, In Nutritional and Dietetic Foods, International Textbook Co. Ltd., London, U. K., p. 64.

Carlson, J.J., Eisenmann, J.C., Norman, G.J., and Ortiz, K.A. (2011). Young. Dietary fiber and nutrient density are inversely associated with the Metabolic Syndrome in US adolescents. JADA, 111; 1688.

Chavan, J.K. and Kadam S.S. (1989). Nutritional improvement of cereals by fermentation. CRC in Food Sci. and Nut. 28: 349-399.

Chavan, J.K. and Salunkhe, D.K. (1984). Structure of sorghum grain. In Nutritional and processing quality of sorghum. Pp. 21-31.

Chavan, J.K., Chavan, U.D. and Nagarkar, V.D. (1989). Effect of malting and fermentation on nutritional quality of sorghum. J. Maharashtra Agric. Univ., 14: 246-247.

Chavan, U.D. and Chavan, J.K. (1991). Utilization of malted sorghum, mung bean and black gram in preparation of bhakari. J. Maharashtra Agric. Univ., 16: 141-142.

Chavan, U.D. (1987). "Effect of Fermentation on the Nutritional Quality of Grain Sorghum (*Sorghum bicolour* L. Moench). "M.Sc. (Agric.) Thesis, Mahatma Phule Agriculture University Rahuri, Maharashtra, INDIA. pp.i-95.

Chavan, U.D. and Chavan, J.K. (1992). Utilization of groundnut flour in preparation of sorghum bhakari. J. Maharashtra Agric. Univ., 17 (20) : 346-347.

Chavan, U.D., Chavan, J.K. and Kadam, S.S. (1988). Effect of fermentation on soluble proteins and *in-vitro* protein digestibility of sorghum, green gram and sorghum + green gram blends. J. Food Sci. 53 : 1574-1575.

Chevassua-Agnes, S., Favier, J.C. and Joseph, A. (1976). Traditional technology and nutritive value of sorghum beer from camaoon (in French) Cah. Nutr. Diet. 11: 89-104.

Committee on Micronutrient Deficiencies, Prevention of Micronutrient Deficiencies (1998). Tools for Policymakers and Public Health Workers. (CP Howson, ET Kennedy and A Horwitz, Eds), National Academy Press. Washington, D.C.

Desikachar, H.S.R. (1977). Processing of sorghum and millets for versatile food uses in India. In Sorghum and millets for human food. Pp. 41-45. Symp. Proc. IACC, Vienna, 1976. Tropical Products Institute, London.

FAO/WHO (1998). Expert Consultation on Human Vitamin and Mineral Requirements.

FAO/WHO/UNU (2001). Report of a Joint Expert Consultation on Human Energy Requirements.

Food and Nutrition Board. (2001). Dietary Reference Intakes for Vitamin A, Vitamin K, Arsenic, Boron, Chromium, Copper, iodine, Iron, Manganese, Molybdenum, Nickel, Silicon, Vanadium and Zinc. National Academy Press, Washington, D.C.

Hahn, R.R. (1969). Dry milling of grain sorghum. Cereal Sci. Today. 14: 234-237.

http://www.cdc.gov/nchs/nhanes/nhanes2005-2006/nhanes05_06.htm

http://www.cnpp.usda.gov/publications/dietaryguidelines/2010/dgac/report/d-5-carbohydrates.pdf

http://www.fda.gov/Food/GuidanceComplianceRegulatoryInformation/GuidanceDocuments/FoodLabelingNutrition/FoodLabelingGuide/ucm064919.htm

Karim, A. and Rooney, L.W. (1972). Characterization of pentosans in sorghum grain. J. Food Sci. 37: 369-371.

Kazanas, N. and Fields, M.L. (1981). Nutritional improvement of sorghum by fermentation. J. Food Sci. 46: 819-821.

Miller, O.H. and Burns, E.E. (1970). Starch characteristics of selected grain sorghum as related to human food. J. Food Sci. 35: 666-668.

Murty, D.S. and Subramanian, V. (1982). Sorghum roti. I. Traditional methods of consumption and standard procedures for evaluation. ICRISAT, 1982, Proc, Intl. Symp. On Sorghum grain quality. 28-31 October, 1981, Patancheru, A.P. India. Pp. 73-78.

Murty, D.S., Patil, H.D., Prasad Rao, K.E. and House, L.R. (1982). A note on screening the Indian sorghum collection for popping quality. J. Food Sci. Technol. 19: 79-80.

Novellie, L. (1977). Beverages from sorghum and millets in "Proc. of Symp. On Sorghum and Millets for Human Food". International Association for Cereal Chemistry. Symposium. Vienna. 1976. Tropical Products Institute, London. Pp. 73-77.

Odunfa, S.A., Nordsrom, J. and Adeniran, S.A. (1994). Development of starter cultures for nutrient enrichment of ogia West African fermented cereal gruel. Report submitted to HBVC research grants program. USAID, Washington, U.S.A.

Patil, U.S, Dalvi, U.S. and Chavan, J.K. (2003). Studies on the production of starch from mould infected discolored sorghum. J. Food Sci. Technol. 40: 115-117.

Raghavandra Rao, S.N. and Desikachar, H.S.R. (1964). Pearling as a method of refining jower and wheat and its effect on their chemical composition. J. Food Sci. Technol. 1: 40-42.

Raghavandra Rao, S.N., Malleshi, N.G., Sreedharamurty, S., Viraktamath, C.S. and Desikachar, H.S.R. (1979). Characteristics of roti, dosa and vermicelli from maize, sorghum and bajra. J. Food Sci. Technol. 16: 21-24.

Reichert, R.D. and Young, C.G. (1976). Dehulling cereal grains and grain legumes for developing countries. I. Quantities comparison between attrition and abrasive type mills. Cereal Chem. 53: 829-839.

Report of the Dietary Guidelines Advisory Committee on the Dietary Guidelines for Americans (2010). USDA.

Salunkhe, D.K., Kadam, S.S. and Chavan, J.K. (1977). Nutritional quality of proteins in grain sorghum. Qual. Plant. Pl. Foods Human Nutr. 27: 187-205.

Schrimshaw, N.S. (1974). Nutritional requirements for high protein foods, Proc. Int. Symp. Protein Foods and Concentrates, Central Food Technol. Res. Institute, Mysore, India, pp. 13-18.

Shepherd, A.D. (1974). How grain structure influences sorghum quality, presented to 5th E. African Cereals Res. Conf. Malawi.

Sullins, R.D. and Rooney, L.W. (1974). Microscopic evaluation of the digestibility of sorghum lines that differ in endosperm characteristics. Cereal Chem. 51: 134–142.

Sulllins, R.D., Rooney, L.W. and Rosenow, D.T. (1975). The endosperm structure of high lysine sorghum. Crop Sci. 15: 599-600.

USDA (2011). National Nutrient Database for Standard Reference, Release 24.

USDA National Nutrient Database for Standard Reference, Release 24 (2011). Item # 20648.

Wang, Y.D. and Fields, M.L. (1987). Germination of corn and sorghum in the home to improve nutritive value. J. Food Sci. 43: 1113-1115.

Watson, S.A. (1970). Wet milling process and products. In Sorghum production and utilization. J. S. Wall and Ross, W. M. Eds. The AVI Publishing Co. Inc, Westport, Connecticut USA. Pp. 602-626.

WHO/FAO/UNU. (2007). Protein and Amino Acid Requirements in Human Nutrition. Report No. 935.

■■■■

Chapter – 2

Medicinal Properties of Sorghum

Some of the health benefits of sorghum include its ability to prevent certain types of cancer, help control diabetes, offer a dietary option for people with Celiac disease, improve digestive health, build strong bones, promote red blood cell development, and boost energy and fuel production.

Sorghum is the broad term for an entire genus of grasses that are native to tropics and subtropics around the world. While there are more than 30 different species of sorghum, only one is harvested for human consumption, while the others are primarily used as fodder for animals. The important species for humans, *Sorghum bicolour,* is native to Africa, but can now be found all around the world as a staple food product. It is also cultivated in many different countries. Sorghum is primarily used in the production of sorghum molasses, sorghum syrup, and as a grain. Also, it can be used in the production of alcoholic beverages and even biofuels around the world. It is widely considered the fifth most important cereal crop in the world.

The versatility of sorghum, combined with the fact that it is acceptable for people with wheat allergies to eat, makes it extremely important as a staple crop in the world. Furthermore, the vast health benefits associated with sorghum make it a great alternative to other types of grains, grasses, and cereals that are commonly consumed across the globe. Let's take a closer look at the nutrients that sorghum can provide us in our diet.

NUTRITIVE VALUE OF SORGHUM

Sorghum is a powerhouse in terms of nutrients and can provide those wise individuals who add it to their diet with vitamins like niacin, riboflavin, and thiamine, as well as high levels of magnesium, iron, copper, calcium, phosphorous, and potassium, as well as nearly half the daily required intake of protein and a very significant amount of dietary fiber (48% of the recommended intake).

The chemical composition of biofuels has not received adequate attention given that it is an important aspect in the introduction of energy crops. In this study, the ash content and mineral composition (C, N, Al, Ca, Cl, Fe, K, Mg, Na, P, S, and Si) of stems, leaves and reproductive organs of some promising energy crops were determined and compared with the respective recommended thresholds reported in literature. Overall, cynara exhibited the highest ash and mineral contents, which indicate high slagging, fouling and corrosion tendencies. However, cynara also showed the lowest Si content, both in leaves (4.3 g kg^{-1}) and in stems (0.9 g kg^{-1}). Sweet sorghum and giant reed exhibited the highest N content (up to 16 g kg^{-1}), which greatly exceeded the recommended limits in leaves. Importantly, Cl always exceeded the recommended limits (up to 18mgkg^{-1} in cynara), both in stems and in leaves, thus resulting in a major stumbling block for all crops. Several significant correlations among elements were found at a single plant part; conversely these correlations were generally very weak considering different plant components, with the exception of K (r ¼ 0.91**), P (r ¼ 0.94**) and ashes (r ¼ 0.64**). Generally, leaves resulted in a significant deterioration of biofuel quality when compared with stems and flower heads. Therefore, agricultural strategies aimed at reducing the leaf component (e.g. by delaying the harvest) may considerably improve the suitability of biofuels for current combustion plants.

HEALTH BENEFITS FROM SORGHUM

1. Digestive Health

It seems like many healthy foods contain some amount of dietary fiber, which greatly improves the functionality of the digestive system. However, sorghum is one of the best foods out there for dietary fiber. A single serving of sorghum contains 48% of your daily recommended intake of dietary fiber, more than 12 grams. That means that your digestive tract will keep your food moving along rapidly, preventing things like cramping, bloating, constipation, stomach aches, excess gas, and diarrhea. Furthermore, excess amounts of fiber in the body helps to scrape off dangerous cholesterol (LDL), which helps to improve heart health and protect your body from conditions like atherosclerosis, heart attacks, and strokes.

2. Grain Sorghum as an Excellent Source of Dietary Fiber

Whole grain gluten-free sorghum flour is an excellent source of dietary fiber with 6.6g/100g flour [USDA National Nutrient Database for Standard Reference, Release 24 (2011). Item no 20648]. Most Americans of all ages under-consume dietary fiber with an average intake of about 15 grams versus the 25 grams that are recommended (http://www.cdc.gov/ nchs/nhanes/nhanes 2005-2006/nhanes05_06.htm). Concentrated sources of fiber include whole grains, dried peas and beans, vegetables, nuts and fruits. Fiber is intrinsic and intact only in plants. Dietary fibers are non-digestible forms of carbohydrates and lignin. Dietary fiber can be classified as fermentable or non-fermentable---both forms are needed for good health. Along with helping provide satiety, a moderate amount of evidence suggests that dietary fibers from whole food sources protect against cardiovascular disease, obesity, and type 2 diabetes and is essential for optimal digestive health (http://www.cnpp.usda.gov/publications/dietary guidelines/2010/dgac/report/d-5-carbohydrates.pdf). The 2010 Dietary Guidelines for Americans was careful to point out that increasing total grains was not recommended. However increased consumption of the proportion of whole grains to total grains was recommended because of the need to increase whole grain fiber (http://www.cnpp.usda.gov/publications/ dietaryguidelines/ 2010/ dgac/report/d-5-carbohydrates.pdf). In particular, individuals with celiac disease may not consume enough dietary fiber and need gluten-free whole grains such as sorghum in their diets. A recent study with adolescents (Carlson *et al.*, 1688) found that higher dietary fiber intakes, but not low saturated fat or cholesterol intakes, were associated with lower incidence of metabolic syndrome. Since metabolic syndrome in adolescence leads to a higher incidence of metabolic syndrome, type 2 diabetes, and cardiovascular disease in adulthood, teens are a major group in need of education about whole grains and other sources of dietary fiber. The Nutrition Facts panel on food labels requires that the amount of dietary fiber per serving be shown. If a food product contains at least 2.5 g/serving, the label may state the food is a good source of fiber. Based upon an extensive review of the evidence, the FDA recently approved a health claim for whole grains that permits manufacturers to state on foods: "Diets rich in whole grain foods and other plant foods and low in total fat, saturated fat, and cholesterol may reduce the risk of heart disease and some cancers" (http://www.fda.gov/Food/ GuidanceCompliance RegulatoryInformation/GuidanceDocuments/Food LabelingNutrition/FoodLabelingGuide/ucm064919.htm). In order to use the claim, the serving size must contain a specified amount of dietary fiber per serving. Four serving sizes are specified ranging from a 35 gram serving with 1.7 grams of fiber to a 55 gram serving with 3.0 grams of dietary fiber (http://www.fda.gov/Food/Guidance ComplianceRegulatoryInformation/ GuidanceDocuments/FoodLabelingNutrition/FoodLabelingGuide/ucm064919.ht m). Thus whole grain sorghum is an excellent source of dietary fiber.

3. Cancer Prevention

The bran layer of the sorghum grains contain important antioxidants that are not found in many other types of food. These antioxidants have been directly connected to a reduced chance of developing various types of cancer, including esophageal cancer, particularly in comparison to people who regularly eat wheat and corn. Antioxidants are the beneficial compounds that neutralize and eliminate free radicals in the body, which often cause healthy cells in the body to mutate into cancer cells.

ANTHOCYANIN EXTRACTION AND QUANTIFICATION FROM SORGHUM

Acidified methanol resulted in significantly higher values for the mono and total anthocyanins than aqueous acetone (**Table 1**). The total anthocyanins extracted by acidified methanol extracts were on average 59% higher than aqueous acetone extracts. Several authors reported that aqueous acetone was better than various alcoholic solvents for fruit procyanidins, anthocyanins and other phenols (Garcia-Viguera et al., 1998; Kallithraka et al., 1995). However, more recently Lu and Foo (2001) observed significant anthocyanin interaction with aqueous acetone to form pyrano-anthocyanidins which significantly lowered quantities of detectable anthocyanins. This reaction was significantly affected by time of anthocyanin–acetone interaction and temperature. The authors did not observe such anthocyanin solvent reactions in acidified methanol, and subsequently proposed acidified methanol as a better solvent. The apparent solvent participation in new product formations with extracted anthocyanins make a direct comparison of actual extraction power of aqueous acetone versus acidified methanol on sorghum anthocyanins difficult. However, since acidified methanol preserves the extracted anthocyanins in their original form better, it should be the solvent of choice for quantification and analysis of anthocyanins. Additional data based on the antioxidant activities of the extracts, and their HPLC profiles may give insight on actual extraction efficiency of the two solvents. This data is discussed in subsequent sections. Sorghum brans had on average three to four times the levels of anthocyanins in grains (**Table 1**). Sorghum anthocyanins are readily concentrated by decortication. Monomeric anthocyanins contributed an average of 30–50% of the anthocyanins in the sorghums. This implies the larger part of sorghum anthocyanins are polymerized or complexed with other compounds and are not readily separated by the extraction conditions used. Such complexed anthocyanins usually have better colour stability in solution than their monomeric constituents (Brouillard, Wigand, Dangles, and Cheminat, 1991; Francis, 1989; Timberlake and Bridle, 1980).

Table 1: Anthocyanin levels[A] in black sorghum brans extracted with Methanol–HCl and acetone-aqueous

Sample Solvent	Monomeric				Total			
	Grain		Bran		Grain		Bran	
	Methanol[B]	Acetone[C]	Methanol[B]	Acetone[C]	Methanol[B]	Acetone[C]	Methanol[B]	Acetone[C]
1998	0.3	nd[D]	2.3	0.3	1.0	0.6	5.3	2.1
1999	1.5	0.5	4.6	1.7	2.7	1.1	10.4	4.8
2000	1.0	0.5	4.2	1.5	2.3	1.4	8.9	4.0
2001	1.0	0.5	4.2	1.5	2.3	1.4	8.9	4.0
2002	1.2	0.4	3.7	1.7	2.8	1.1	11.0	5.2
Mean	0.9[a]	0.5[b]	3.3[a]	1.1[b]	2.0[a]	1.0[b]	8.1[a]	3.4[b]
LSD	0.1	0.2	0.2	0.4	0.1	0.2	0.3	0.4

[A] mg/g Luteolinidin equivalents (DM basis), determined pH differential method. Samples were extracted for 2 h at room temperature and equilibrated for 24 h at -20°C before analysis. [B]Acidified methanol (1% HCl). [C]70% aqueous acetone. [D]nd, Not detected at significant levels. All comparisons between the two solvents were significant at ($\alpha = 0.05$).

Seasonal variations in anthocyanin contents were significant. However, since no specific environmental parameters were tested or controlled for, the factors that contributed to the observed variability cannot be pinpointed. However, the 1998 and 2000 grains (which had the lowest anthocyanin contents) had wrinkled, poorly formed seeds. This suggests significant environmental stress during seed development. Such stress may have negatively affected anthocyanin biosynthesis. For example, Dube, Bharti, and Laloraya (1992) showed that ionic stress may inhibit light-induced anthocyanin synthesis in sorghum. Sene, Dore and Gallet (2001) also reported that nitrogen nutrition and environmental factors that promote growth and grain yield also enhance phenol synthesis in sorghum. It is important to conduct additional studies to determine which specific environmental factors influence the anthocyanin contents of the sorghum. This will help maximize anthocyanin production in sorghum.

Table 2: Molar properties and antioxidant activity of anthocyanin standards

Standard	λmax (pH 1)[a]	ε (pH 1)[b]	Antioxidant activity[c]	Number of OH groups
Luteolinidind	482	31,700±1100	4.81	4
Apigeninidind	468	30,400±500	4.20	3
Peonidin	516	27,200±4100	3.62	4
Pelargonidin	506	28,100±200	3.33	4
Pelargonidin-3, 5-diglcoside	498	28,800±1100	2.88	2
Cyanidin	516	24,800±60	5.35	5
Cyanidin-3-glucoside	512	28,600±800	4.60	4
Cyanidin-3, 5-diglosidec	510	33,900±1700	6.34	3
Cyanidin-3-rutinoside	514	26,100±1700	4.20	4

[a]Maximum absorption wavelength in pH 1 buffer. [b]Extinction coefficient in pH 1 buffer. [c]mol Trolox/mol, ABTS method. [d]Compounds identified in sorghum. LSD for antioxidant activity = 0.11 ($\alpha > 0{:}05$).

HPLC ANALYSIS

The absorption characteristics and other properties of the anthocyanin standards analysed are shown in **Table 2**. The 3-deoxyanthocyanidins, luteolinidin and apigeninidin had absorption maximas that were particularly different from those of the other anthocyanidins. These two anthocyanidins are structurally different from the rest of the anthocyanidins which are commonly found in fruits and vegetables since they lack an oxygen molecule at the C-3 position (**Fig. 1**). Their absorption maximas of 468 nm (apigeninidin) and 482 nm (luteolinidin) in pH 1 buffer solution made them appear yellow and orange, respectively. This was in contrast with the other anthocyanidins which were all reddish at pH 1. At near neutral pH (in methanol), the 3-deoxyanthocyanidins appeared yellowish orange and reddish orange for the apigeninidin and luteolinidin, respectively. The rest of the anthocyanins ranged from red to dark blue in colour at neutral pH.

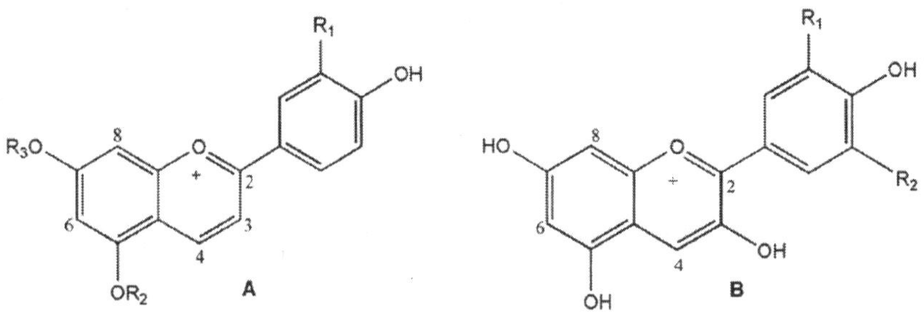

R_1 = H, R_2 = H, R_3 = H: apigeninidin
R_1 = H, R_2 = Glc, R_3 = H: apigeninidin-5-glucoside
R_1 = H, R_2 = H, R_3 = CH_3: 7-O-methyl apigeninidin
R_1 = OH, R_2 = H, R_3 = H: luteolinidin
R_1 = OH, R_2 = Glc, R_3 = H: luteolinidin-5-glucoside
R_1 = OH, R_2 = CH_3, R_3 = H; 5-methoxyluteolinidin

R1 = OH, R2 = H: cyanidin
R_1 = R_2 = H: pelargonidin
R1 = OCH_3, R_2 = H: peonidin
R_1 = R_2 = OCH_3: malvidin
R_1 = R_2 = OH: delphinidin
R_1 = OCH_3, R_2 = OH: petunidin

Figure 1: Chemical structure of some common anthocyanidins compared to 3-deoxyanthocyanidins, luteolinidin and apigeninidin.

The lack of oxygen on C-3 of the 3-deoxyanthocyanidins is thought to accord them greater stability in solution compared to the other anthocyanidins (Iacobucci and Sweeny, 1983; Sweeny and Iacobucci, 1981; Timberlake and Bridle, 1980). For example, Timberlake and Bridle (1980) reported that apigeninidin was stable in pH 2.8 solution for up to 1 year at room temperature and laboratory light, whereas cyanidin degraded within a few hours under similar conditions. Hence the 3-deoxyanthocyanidins may have an advantage over the other anthocyanidins in food applications. The HPLC profiles for the sorghums and their brans were all similar, hence only one sample (2001 bran) profile is shown for illustration purposes (**Figures 2 and 3**). Among the anthocyanin standards used, only the 3-deoxyanthocyanidins (apigeninidin and luteolinidin) were identified in acidified methanol sorghum extracts (**Figure 2** and **Table 3**). Other HPLC peaks, including one major, and five minor ones from the black sorghums, could not be identified due to lack of appropriate standards. The unknown peaks 2 and 3 (**Figure 2**) had spectral characteristics similar to apigeninidin. They were tentatively identified as glycosides of apigeninidin, based on their elution times and spectral characteristics. The other unidentified peaks were all spectrally similar to luteolinidin, suggesting they may be structurally related.

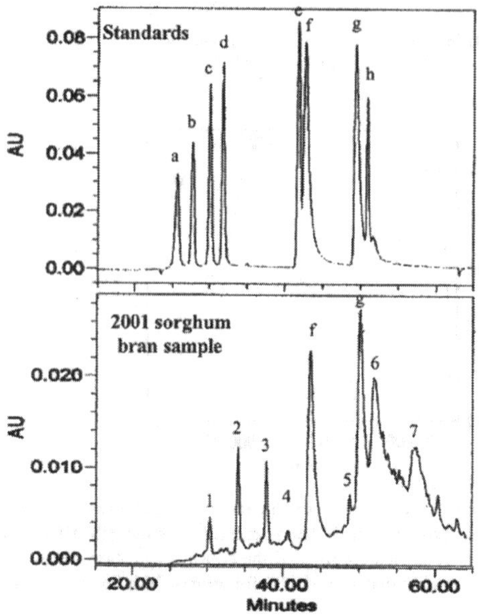

Figure 2: HPLC profiles of acidified methanol (1% HCl) anthocyanin extracts from 2001 black sorghum bran. Detection was at 480 nm: (a) cyanidin-3,5-diglucoside; (b) pelargonidin-3,5-diglucoside; (c) cyanidin-3-glucoside; (d) cyanidin-3-rutinoside; (e) cyanidin; (f) luteolinidin; (g) apigeninidin; (h) peonidin. Peaks 2 and 3 were tentatively identified as glycosides of apigeninidin. Peaks 1 and 4–7 were spectrally similar, and possibly related to luteolinidin. Seventy percent of aqueous acetone extracts did not give significant peaks at this wavelength.

Aqueous acetone extracts produced very low peaks at the 480 nm wavelength used for anthocyanin detection. However, the aqueous acetone extracts had four major distinct peaks, 2 of which had elution charactersistics corresponding to luteolinidin and apigeninidin, but with absorption maximas at between 370 and 380 nm, instead of the typical 470–490 nm for these anthocyanidins in the HPLC solvents (**Figures 3** and **4**). The typical absorption peaks for luteolinidin and apigeninidin were still present but at much reduced intensity (**Figure 4**). Apparently, the anthocyanin molecules undergo significant structural modification in aqueous acetone, a phenomenon that did not occur in acidified methanol. This observation agrees with that of Lu and Foo (2001), who were able to confirm formation of pyrano-anthocyanins from anthocyanins through oxidative addition mediated by acetone. These compounds may be responsible for the observed spectra of the aqueous acetone extracts. Thus aqueous acetone is not an appropriate solvent for extracting sorghum anthocyanins for HPLC characterization. Apigeninidin and luteolinidin represented an average of 49% of the total peak areas, whereas the major unknown peak (peak number 6) represented about 25% of the total black sorghum anthocyanins (**Table 3** and **Figure 2**).

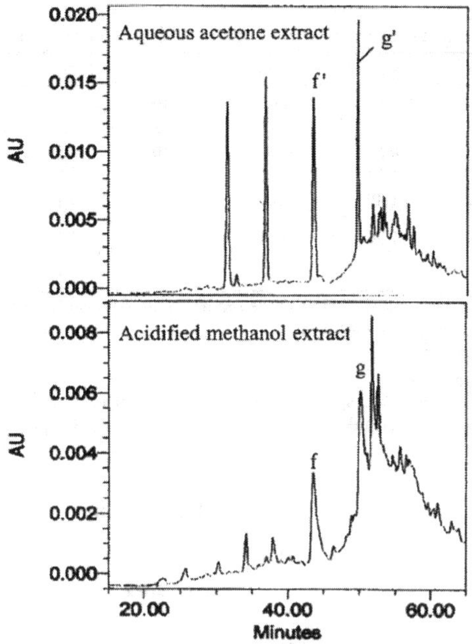

Figure 3: HPLC anthocyanin profiles of 2001 black sorghum bran extracts monitored at 375 nm. Note the increased intensity of resolved peaks for 70% aqueous acetone extracts relative to acidified methanol (1% HCl) extracts at this wavelength. Peaks 'f' and 'g' correspond to acetone-modified luteolinidin and apigeninidin, respectively (see Figure 4). Peaks 'f' and 'g' are luteolinidin and pigeninidin, respectively. Samples were extracted for 2 h and equilibrated at -20°C for at least 24 h before analysis.

Table 3: Anthocyanin contents of black sorghum brans determined by HPLC

Year	Luteo[a]	Apig[b]	1[c]	2[d]	3[d]	4[c]	5[c]	6[c]	7[c]	Total
1998	0.89	0.63	0.10	0.19	0.20	Trace	0.04	0.47	0.26	2.78
1999	1.77	0.43	0.08	0.19	0.14	Trace	0.22	0.09	0.21	6.12
2000	0.53	0.57	0.10	0.21	0.14	0.10	Trace	0.15	Trace	1.79
2001	1.34	1.42	0.10	0.26	0.29	0.05	Trace	1.05	0.58	5.35
2002	1.49	1.26	0.27	0.25	0.23	Trace	0.11	1.55	0.63	5.96
Mean	1.20	0.86	0.13	0.22	0.20	0.03	0.07	12.6	0.34	4.40
CV	1.4	1.8	0.9	1.5	1.2	0.8	0.9	1.4	4.5	2.3

All unknowns and totals expressed as luteolinidin equivalents. Units are mg/g DM basis. [a]Luteolinidin. [b]Apigeninidin. [c]Unidentified, had spectral characteristics similar to luteolinidin. [d]Tentatively identified as glycosides of apigeninidin. Samples were extracted with 1% HCl in methanol. CV, coefficient of variation.

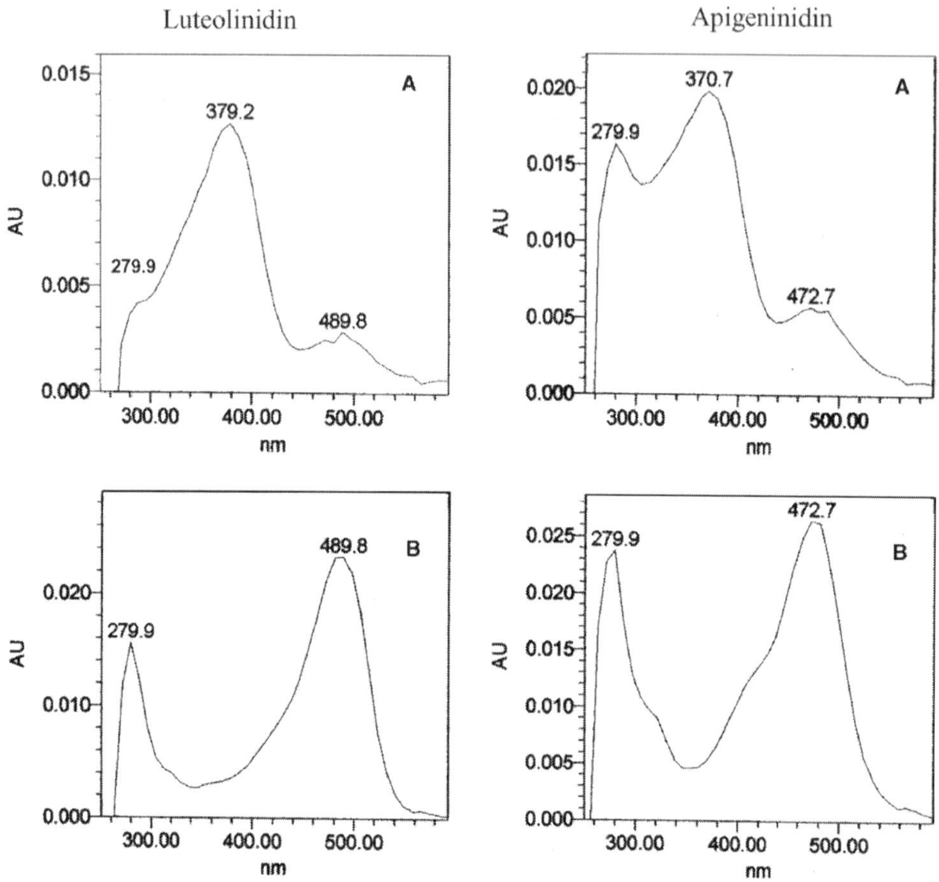

Figure 4: Spectral characteristics of peaks corresponding to solvent modified (A) and unmodified (B) luteolinidin and apigeninidin isolated from 2001 black sorghum bran by HPLC. Spectra 'A' were obtained from 70% aqueous acetone extracts and 'B' from acidified methanol (1% HCl) extracts. Note the prominent presence of a third peak, not typical for anthocyanins, and the diminished intensity of the typical absorption maxima of the anthocyanidins in the aqueous acetone extracts. Samples were extracted for 2 h and equilibrated at < -20°C for at least 24 h before analysis.

Gous (1989) also identified luteolinidin and apigeninidin as the major anthocyanins in a black sorghum variety. Glycosylated forms of these anthocyanins were also identified in sorghums by Nip and Burns (1969, 1971). Unlike most anthocyanins in fruits and vegetable that are naturally found as glycosides, the 3-deoxyanthocyanidins do exist in nature as aglycones (Clifford, 2000; Stafford, 1965). Hence their abundance in the sorghum extracts may not necessarily mean degradation during extraction. The total anthocyanin contents measured by HPLC were strongly correlated with those measured by the pH differential method (R^2 = 0:98). The simpler pH differential method gives a good estimation of anthocyanin content of black sorghum.

Comparison of black sorghum with commercial anthocyanin sources

Black sorghum brans were a very good source of anthocyanins (4.0–9.8 mg/g) relative to the commercial sources currently available (0.2–10 mg/g) (**Table 4**). This coupled with the fact that they possess mostly the relatively stable 3-deoxyanthocyanidins gives them an edge over the fruits and vegetables as a source of natural anthocyanins. The sorghum has an additional advantage in terms of storage stability relative to the fruits and vegetables. Sorghum is harvested at low moisture (13–15%) compared to fruits and vegetables which normally have >80% moisture and require significant time and energy for drying to improve their storage stability. The black sorghum can be a competitive source of anthocyanins.

Table 4: Anthocyanin content of sorghum brans relative to common fruits and vegetables

Commodity	Content[a] (mg/g)
Black sorghum bran	4.0-9.8
Blueberry	0.2-5.0[b,c]
Cranberry	0.6-2.0[c]
Black currant	0.3-4.0[b,c]
Red grapes	0.3-7.5[d]
Black raspberry	1.7-4.3[b,c]
Red raspberry	0.1-0.6[b,c]
Strawberry	0.2-0.9[c,f]
Elderberry	2.0-10.0[g]
Red cabbage	0.3-0.9[b,d]

Sorghum samples extracted in 1% HCl in methanol. [a]Fresh weight. [b]Wang et al. (1997). [c]Clifford (2000). [d]Bridle and Timberlake (1997). [e]Torre and Barrit (1997). [f]Garcia-Viguera et al. (1998). [g]Bronnun-Hansen et al. (1985).

ANTIOXIDANT ACTIVITY OF BLACK SORGHUM

Antioxidant activities of samples extracted in acidified methanol and aqueous acetone were compared (**Figure 5**). Samples extracted in acidified methanol had significantly higher antioxidant activity than those extracted in aqueous acetone. This implies that the acidified methanol is a more powerful solvent than aqueous acetone at extracting black sorghum antioxidants. The 3-deoxyanthocyanidins found in the black sorghum had antioxidant activities that were similar to the anthocyanins found in fruits and vegetables (**Table 2**). Additionally, a good correlation (R^2 = 0:94) was observed between total anthocyanin levels and antioxidant activity of the sorghum samples (**Figure 6**), confirming the major contribution of anthocyanins to black sorghum antioxidant activity. The sorghum grains and brans had significantly higher phenols and antioxidant activity than other cereals (**Table 5**). Cereal brans from wheat, barley, buckwheat, and rice, among others, are promoted as good sources of antioxidants (Emmons and Peterson, 1999; Haber, 2002; Quettier-Deleu et al., 2000; Yu et al., 2002; Zielinski and Kozlowska, 2000). Such brans are sold in the market for use in fortified baked products. Black sorghum bran offers a major advantage in terms of antioxidant value per unit weight. The sorghum bran can be used as a high value source of antioxidants at lower quantities than other cereal brans, or used at similar quantities to provide higher antioxidant activity in products. Additionally black sorghum bran imparts a natural dark appealing colour normally associated with 'healthy' baked goods, and was shown to produce acceptable quality bread (Gordon, 2001) and cookies (Mitre-Dieste, Gordon, Awika, Suhendro, and Rooney, 2000) at levels of up to 15% and 50%, respectively. Since the black sorghum had no condensed tannins, any perceived nutritional concerns related to tannins are eliminated.

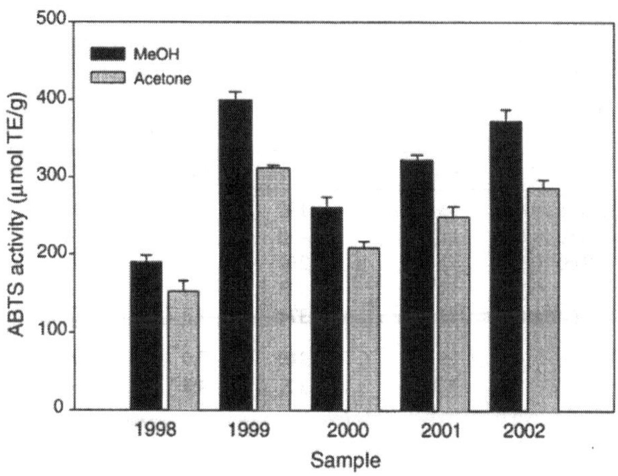

Figure 5: Comparison of antioxidant activities of black sorghum brans extracted in acidified methanol (1% HCl) and 70% aqueous acetone. Samples were extracted for 2 h and equilibrated at <-20°C for at least 24 h before analysis. Differences within each sorghum were significant (α = 0:05).

In summary, black sorghum brans were a very good source of anthocyanins (4.0–9.8 mg/g) compared to commercial sources currently available (0.2–10 mg/g). The black sorghum anthocyanins were composed largely of the 3-deoxyanthocyanidins, which are more stable than the anthocyanins mostly found in fruits and vegetables used currently as commercial sources of anthocyanins. This should make the sorghum anthocyanins competitive as a source of natural food colour. The antioxidant activity of the sorghum anthocyanins were similar to those of the anthocyanins found in fruits and vegetables, hence they may offer similar health benefits. The black sorghum brans are superior to other cereal brans as a source of antioxidants. They may provide more health benefits when used in cereal based foods than the current commercial brans.

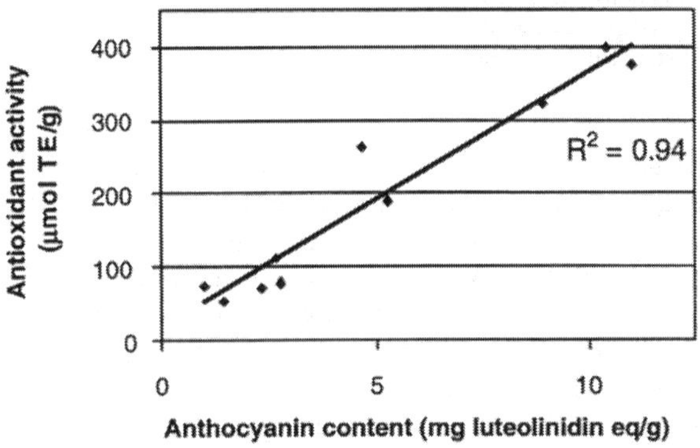

Figure 6: Correlation between anthocyanin content and antioxidant activity of black sorghums and their brans. The ABTS method was used for antioxidant assay. Anthocyanins were determined by the different method. Acidified methanol (1% HCl) was the extracting solvent.

Table 5: ABTS Antioxidant activity and phenol levels among black sorghums and other cereals

Sample	Activity (µmol TE/g)		Phenol (mg/g)	
	Grain	**Bran**	**Grain**	**Bran**
Black 1998	75	190	5.6	18.1
Black 1999	112	400	10.0	32.0
Black 2000	52	261	5.5	19.1
Black 2001	72	323	6.4	26.0
Black 2002	78	374	7.3	35.6
Mean	78	310	7.0	26.2
CV	2.8	3.5	4.0	5.0
Wheat	<0.1–2.0[a;b]	28[c]–34	0.4–0.9[b]	3.0
Barley	9.11[a]	-	-	-
Rye	4[a]	-	-	-
Oats	3[a]	5[c]	0.2-0.3[d]	0.2-0.3[d]
Buckwheat	25[a]	-	3.2[f]	3.3[e]

Sorghum samples extracted with 1% HCl in methanol. [a]Zielinski and Kozlowska (2000). [b]Yu et al. (2002). [c]Haber (2002). [d]Emmons and Peterson (1999). [e]Quettier-Deleu et al. (2000). [f]Estimated from Quettier-Deleu et al. (2000).

Cancer preventing compounds present in sorghum

Sorghum is a good source of phenolic compounds (Hahn, Faubion, and Rooney, 1983). Protocatechuic, *p*-hydroxybenzoic, caffeic, *p*-coumaric and ferulic acids are found in the free and bound forms. However, cinnamic and vanillic acids are found in the free and/or bound forms only in some sorghum varieties. Gallic acid is present only in the bound form (Hahn et al., 1983). Recently, four phenolic acids (caffeic, *p*-coumaric, ferulic and sinapic acids) were detected in all pearling fine and pearled kernel fractions of two sorghum genotypes namely white (PR6E6) and red (PR6E14) (Luthria and Liu, 2013). Sorghum tannins readily associate with sorghum proteins (Shahidi and Naczk, 2004). Only varieties with a pigmented testa have condensed tannins. Sorghum tannins are excellent antioxidants, which slow hydrolysis in foods, produce naturally dark-coloured products and increase the dietary fiber levels of food products (Dykes and Rooney, 2006). The proanthocyanidins in sorghums are the B-type with (–)-epicatechin as extension units and catechin as terminal units (Gu *et al.,* 2002, 2003; Gupta and Haslam, 1978). The anthocyanins are the major class of flavonoids studied in sorghum. Unlike common anthocyanins, sorghum anthocyanins are unique since they do not contain the hydroxyl group in the 3-position of the C-ring and thus are called 3-deoxyanthocyanins (Dykes and Rooney, 2006). The antioxidant properties of the 3-deoxyanthocyanidins are similar to those of the anthocyanins (Awika, Rooney, and Waniska, 2004a). The two common 3-deoxyanthocyanidins in sorghum are the yellow, apigeninidin, and the orange, luteolinidin (Awika, Rooney, andWaniska, 2004b). Other 3-deoxyanthocyanins identified in sorghum grains include apigeninidin 5-glucoside, luteolinidin 5-glucoside, 5-methoxyluteolinidin, 5-methoxyluteolinidin 7-glucoside, 7-methoxyapigeninidin, 7-methoxyapigeninidin 5-glucoside, 5-methoxyapigeninidin and 7-methoxyluteolinidin (Dykes and Rooney, 2006). Red pericarp sorghums have flavan-4-ol compounds, such as luteoforol and apiforol, which are produced from flavanones (i.e., naringenin and eriodictyol) and may be precursors of sorghum anthocyanidins (Dykes and Rooney, 2006). Other flavonoids isolated and identified in sorghum grains include the flavones apigenin and luteolin, which are predominant in tan-pigmented sorghums. Flavanones, eriodictyol and eriodictyol 5-glucoside have been reported. The flavonol, kaempferol 3-rutinoside-7-glucuronide and the dihydroflavonols taxifolin, and taxifolin 7-glucoside have also been isolated (Dykes and Rooney, 2006). Kil *et al.* (2009) suggested that sorghum could be used as a natural ingredient with biological function for its antioxidant and antimicrobial properties in the food industry. Fernandez *et al.* (2009) reported that some sorghum varieties indigenous to West Africa (Nigeria and Niger) have a yellow endosperm, which is pigmented by carotenoids. The most abundant carotenoids identified are lutein, zeaxanthin and β-carotene (Taylor, Belton, Beta, and Duodu, 2014).

Sorghum (*Sorghum bicolour* M.) is the fifth leading cereal food crop in the world and is particularly important as a human food resource in Asia and Africa

(Rooney and Waniska, 2000). Sorghum is a rich source of various phytochemicals including tannins, phenolic acids, anthocyanins, phytosterols and policosanols (Awika and Rooney, 2004). Tannin content of sorghum grain ranged from 10.0 – 68.0 mg/g (Awika *et al.*, 2004), in which only sorghums containing a pigmented testa had significant amounts of condensed tannins (Dykes *et al.*, 2005). Like other cereals such as wheat, corn and barley, the sorghum phenolic compounds are mostly concentrated in the bran and in bound forms. Ferulic acid is the most abundant bound phenolic acid in sorghum (Hahn *et al.*, 1983), while other phenolic acids including syringic, protocatechuic, caffeic, p-coumaric, and sinapic are also found in sorghum as the more abundant (Waniska *et al.*, 1989). The most common anthocyanins in sorghums are the 3-deoxyanthocyanidins, which include orange luteolinidin and yellow apigeninidin. Both compounds have good potential for use as natural colourants due to their pH stability (Dykes *et al.*, 2005). Recently, a number of studies on health-related benefits of sorghum reported that sorghum has significantly antioxidant activity (Choi *et al.*, 2006), antimicrobial activities (Kil *et al.*, 2009), anticarcinogenic effects (Kwak *et al.*, 2004), and cholesterol-lowering effects (Ha *et al.*, 1998), and can reduce the risk of cardiovascular disease (Cho *et al.*, 2000).

Table 6: Phenolic acids and flavonoids reported in cereal grains.

Compound	Grain
A. Phenolic acid	
1. Hyderoxybenzoic acids	
Gallic	Millet, rice, sorghum
Protocatechuic	Barley, maize, millet, oat, rice, rye, sorghum, wheat
ρ-Hydroxybenzoic	Barley, maize, millet, oat, rice, rye, sorghum, wheat
Gentisic	Millet, sorghum
Salicylic	Barley, sorghum, wheat,
Vanillic	Barley, maize, millet, oat, rice, rye, sorghum, wheat
Syringic	Barley, maize, millet, oat, rice, rye, sorghum, wheat
2. Hydroxycinnamic acids	
Ferulic	Barley, maize, millet, oat, rice, rye, sorghum, wheat
Caffeic	Barley, maize, millet, oat, rice, rye, sorghum, wheat
α-Coumaric	Barley
m-Coumaric	Barley
ρ-Counaric	Barley, maize, millet, oat, rice, rye, sorghum, wheat

Compound	Grain
Cinnamic	Millet, sorghum, wheat
Sinapic	Barley, millet, oat, rice, rye, sorghum
B. Flavonoids	
1. Anthocyanins	
Apigeninidin	Sorghum
Apigeninidin-5-glucoside	Sorghum
Cyanidin	Barley
Cyanidin 3-galactoside	Maize, wheat
Cyanidin 3-glucoside	Barley, maize, rice, rye, wheat
Cyanidin 3-rutinoside	Maize, rice, wheat
Delphinidin	Barley
Delphinidin 3-glucoside	Wheat
Delphinidin 3-rytinoside	Rye, wheat
Luteolinidin	Sorghum
Luteolinidin 5-glucoside	Sorghum
5-Methoxyapigeninidin	Sorghum
7- Methoxyapigeninidin	Sorghum
7- Methoxyapigeninidin 5-glucoside	Sorghum
5-Methoxyluteolinidin	Sorghum
5-Methoxyluteolinidin 7-glucoside	Sorghum
7- Methoxyluteolinidin	Sorghum
Pelargonidin	Barley
Pelargonidin 3-glucoside	Maize
Pelargonidin glucosides	Barley, maize
Peonidin-3-glucoside	Maize, rice, rye, wheat
Petunidin 3-glucoside	Barley, wheat
Petunidin 3-rutinoside	Wheat
2. Flavones	
Apigenin	Millet, oat, sorghum
Apigenin glucosides	Wheat
Glucosylorientin	Millet
Glucosylvitexin	Millet
Luteolin	Millet, oat, sorghum
Isovitexin	Oat
Tricin	Millet, oat, wheat
Vitexin	Millet, oat
3. Flavanones	
Eriodictyol	Sorghum

Compound	Grain
Eriodictyol 5-glucoside	Sorghum
Homoeriodictyol	Oat
Naringenin	Sorghum
4. Flavonols	
Chrysoeriol	Barley
Kaempferol	Maize, oat
Kaempferol 3-rutinoside	Oat
Kaempferol 3-rutinoside-7- glucuronide	Sorghum
Quercetin	Maize, oat
Quercetin 3-rutinoside	Oat
5. Dihydroflavonols	
Taxifolin	Sorghum
Taxifolin 7-glucoside	Sorghum
6. Flavan-4-ols	
Apiforol	Sorghum
Luteoforol	Sorghum
7. Flavanols (monomers/dimers)	
Catechin	Barley, sorghum
Leucocyanidin	Barley, maize
Leucodelphinidin	Barley
Leucopelargonidin	Maize
Procyanidin B-1	Sorghum
Procyanidin B-3	Barley
Prodelphinidin B-3	Barley

Phenolic acids are derivatives of benzoic and cinnamic acids (**Figure 7**) and are present in all cereals (**Table 6**). Other flavonoids found in fruits vegetables are also reported in cereals. For example the flavone apigenin a compound found in parsley and celery, is also reported in millet, oat and sorghum (**Table 7**). Flavonoids are reported to have antioxidant, anticancer, anti-allergic, anti-inflammatory, anticarcinogenic and gastroprotective properties. There are two classes of phenolic acids; hydroxybenzoic acids and hydroxycinnamic acids. Hydroxybenzoic acids include gallic, ρ-hydroxybenzoic, vanillic, syringic and protocatechuic acids. The hydroxycinnamic acids have a C_6– C_3 structure and include coumaric, caffeic, ferulic and sinapic acids. The phenolic acids reported in cereals occur in both free and bound forms. Sorghum and millet have the widest variety of phenolic acids. Free phenolic acids are located in the outer layer of the pericarp and are extracted using organic solvents. Bound phenolic acids are esterified to cell walls; acid or base hydrolysis is required to release these bound compounds from the cell matrix. The major phenolic acids in cereals are ferulic and ρ-coumaric acids. Phenolic acid levels vary among cereals; their brans concwntrate these compounds threefold (**Table 7**).

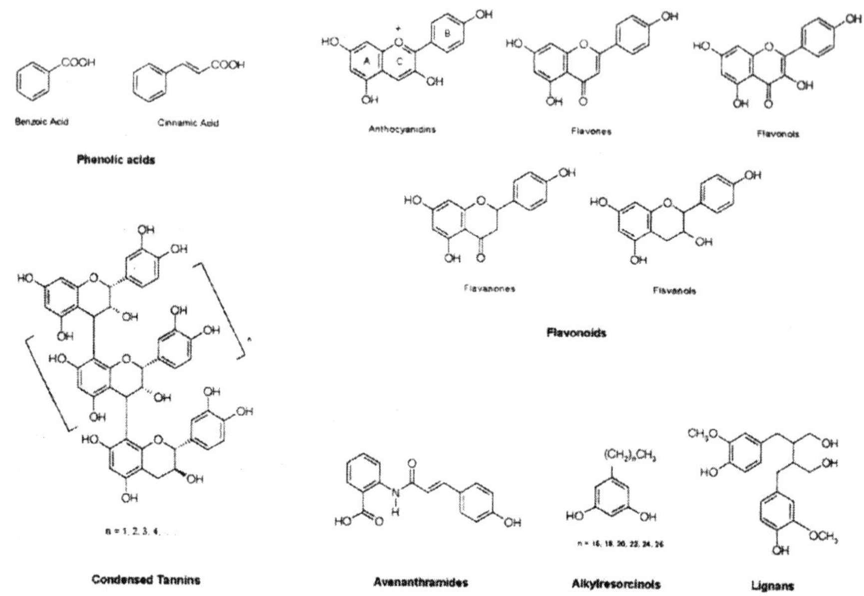

Figure 7: Chemical structure of classes of phenolic compounds in cereal grains.

Table 7: Phenolic acid content in cereal grains

Sample	Amount (μg/g)
A. Whole grains	
Barley	450-1346
Finger millet	612
Foxtail millet	3907
Maize	601
Oat	472
Pearl millet	1478
Rice	197-376
Rye	1362-1366
Sorghum	385-746
Wheat	1342
B. Brans	
Oat	651
Rye	4190
Wheat	4527

Figures 8 and 9 compare phenol and in vitro antioxidant activity levels of a wide array of cereal grains. The Folin-Ciocalteu and the 2, 2'-azinobis (3-ethyl-benzothiazoline-6-sulfonic acid (ABTS) assays were used. respectively using 1% HCI in methanol as the solvent. In general, tannin-containing grains (i.e., sorghum) and pigmented cereal grains had the highest levels of phenols and antioxidant activity in each grain category. Tannin sorghums and black rice had the highest levels of phenols and antioxidant activity whereas non-pigmented cereals (i.e., white rice, wheat, and waxy barley) had the lowest levels. These results suggest that condensed tannins and pigment-contributing compounds such as the anthocyanins increase phenols and antioxidant activity. There is a high correlation between total phenols and antioxidant activity (R^2 = 0.96), which suggests that the antioxidant activity is contributed by the phenolic compounds. In vitro methods used to measure antioxidant activity (i.e., ABTS, DPPH, and ORAC) do not give information about the bioavailability or metabolism of these compounds in biological systems. However, these methods are useful to screen and compare antioxidant activity levels among a wide variety of samples. To date, reports on human health benefits of cereal phenols are limited and, therefore, more research is needed in this area.

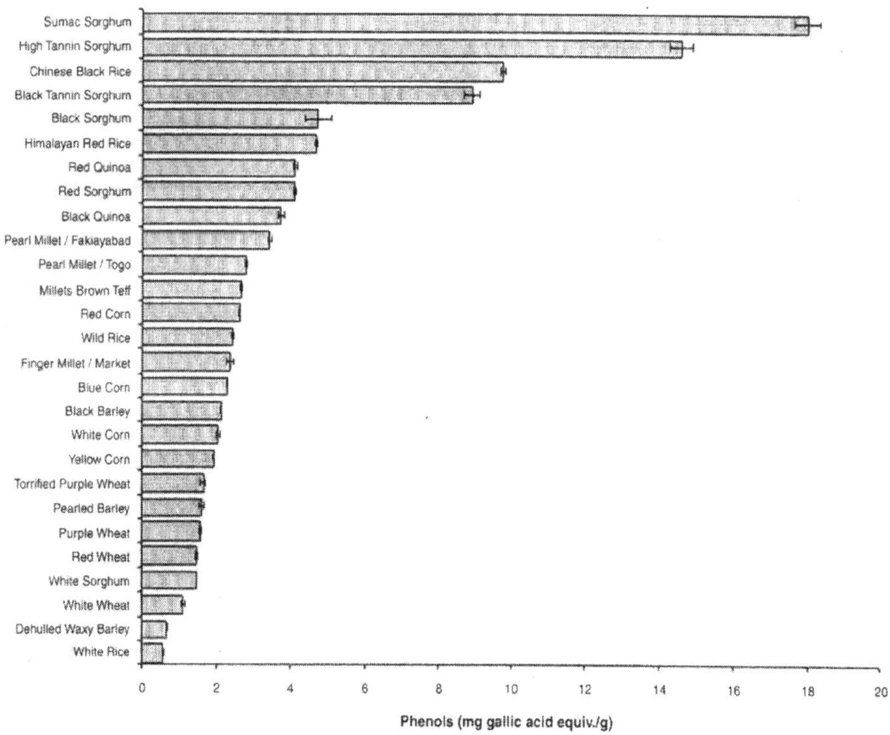

Figure 8: Total phenol levels of cereal grains (Adapted from Guajardo-Flores *et al.*, 2006).

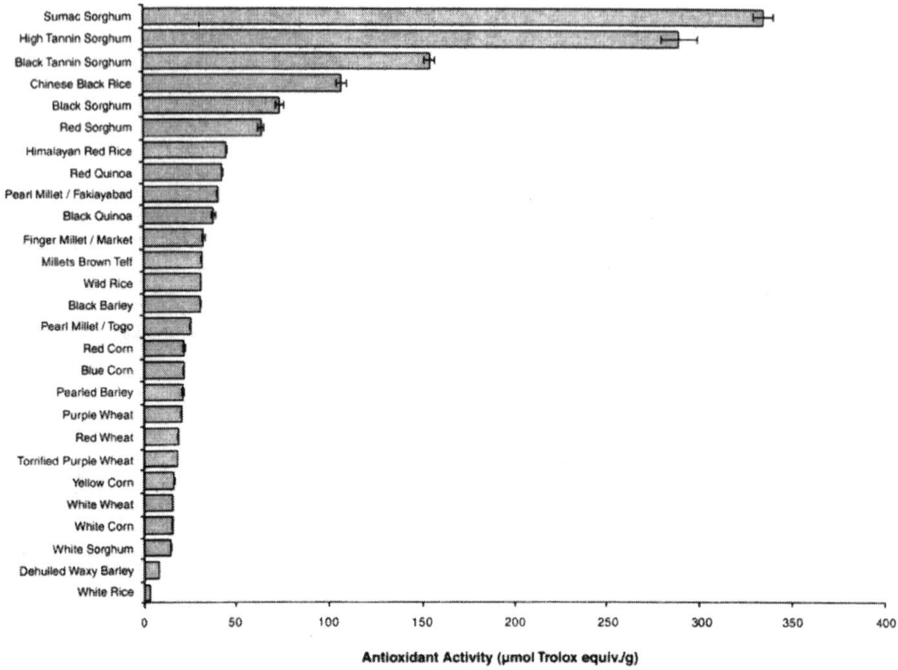

Figure 9: Antioxidant activity (ABTS) levels of cereal grains (Adapted from Guajardo-Flores *et al.*, 2006).

Sorghums containing condensed tannins have consistently shown the highest antioxidant activity *in vitro*, and they approach or exceed the antioxidant levels of fruits and vegetables (**Table 8**). Sorghums containing condensed tannins dominate production of grains in hot, humid regions of Africa because they have significantly improved resistance to grain molds and birds, which allow for their successful production. Tannin sorghums contain a pigmented testa, which contributes astringency during grain maturation and causes birds to utilize other food sources. When other grains are unavailable, birds consume the tannin sorghums. These tannin sorghums are grown extensively in east and southern Africa.

The condensed tannins cause depressed feed efficiency because they slow down and decrease digestion of the grain components. Thus, the tannin sorghums are discriminated against in sorghum markets. Very little tannin sorghums are grown in the United States since they are discounted in the grain markets. In Africa, the tannin sorghums are used in a wide variety of traditional foods including beer, porridges, unleavened breads, and other products. Special tannin sorghums are consumed when farmers are doing field work. The tannin sorghum porridges stay with the person longer and are said to be more satisfying, probably because of the reduced rate of digestion. The estimated glycemic index of ground whole sorghums with tannins is significantly lower than sorghums without tannins.

Table 8: Antioxidant activity (ORAC) of sorghum grains and brans compared to common fruits and vegetables.

Commodity	ORAC (μmol TE/g, dry wt.)
Tannin sorghum (grain)	868
Tannin sorghum (bran)	3124
Black sorghum (grain)	219
Black sorghum (bran)	1008
Red sorghum (grain)	140
Red sorghum (bran)	710
White sorghum (grain)	22
White sorghum (bran)	64
Blueberry, low bush	842
Strawberry	402
Plum	495
Watermelon	18
Apple red delicious	295
Orange navel	137
Broccoli	173
Carrot	108
Onion red	93
Sweet pepper green	105
Radishes	217
Potatoes russet	63

Adapted from Dykes and Rooney (2006).

Sorghum contains unique anthocyanins called 3-deoxyanthocyanins, which lack the hydroxyl group in the 3 position of the C-ring (**Figure 10**).This feature is believed to increase their stability at high pH compared to common anthocyanins, which could make them good natural food colourants. The two main sorghum 3-deoxyanthocyanidins are the yellow apigeninidin and the orange luteolinidin. Sorghum with a black pericarp has higher levels of 3-deoxyanthocyanins than red sorghums. Decortication of black sorghum grain to produce bran concentrates anthocyanin levels almost seven fold. For example the grain and bran of Tx430 black have anthocyanin levels of 944 (**Table 9**) and 6.695 g/g respectively (unpublished data).

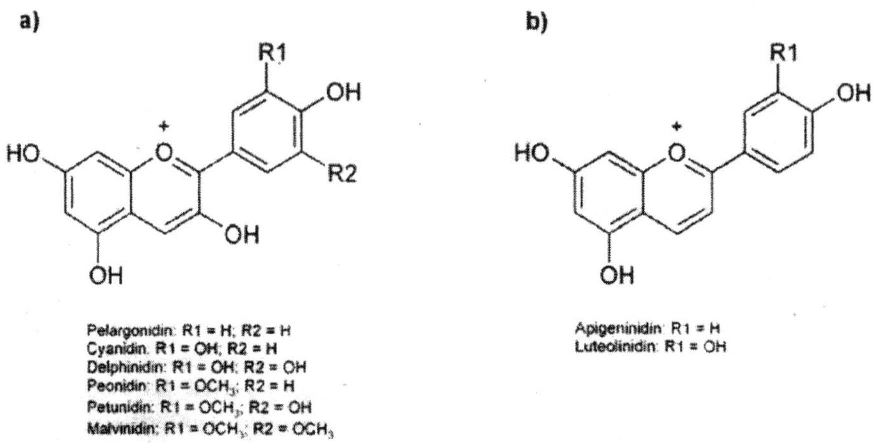

Figure 10: Chemical structure of A, the six common anthocyanidins and B, the 3-deoxyanthocyanidins.

Table 9. Anthocyanin content of Pigmented Cereal Grains

Sample	Amount (μg/g)
Blue barley	4
Maize	
Pink	93
Red	558
Blue	225
Purple	965
Black rice	2283
Black sorghum	944
Wheat	
Blue	106-153
Purple	13-139

Sorghum is a good source of phenolic compounds with a variety of genetically dependent types and levels including phenolic acids, flavonoids, and condensed tannins. Most sorghum does not contain condensed tannins, but all contain phenolic acids. Pigmented sorghums contain unique anthocyanins that could be potential food colourants. Some sorghum has a prominent pigmented testa that contains condensed tannins composed of flavan-3-ols with variable length. Flavan-3-ols of up to 8–10 units have been separated and quantitatively analyzed. These tannin sorghums are excellent antioxidants, which slow hydrolysis in foods, produce naturally darkcoloured products and increase the dietary fiber levels of food products. Sorghums have high concentration of 3-deoxyanthocyanins (i.e. luteolinidin and apigenidin) that give stable pigments at high pH. Pigmented and tannin sorghum varieties have high antioxidant levels that are comparable to fruits and vegetables. Finger millet has tannins in some varieties that contain a red testa. There are limited data on the phenolic compounds in millets; only phenolic acids and flavones have been identified.

CONTRIBUTION OF SORGHUM/MILLET PHENOLS TO HEALTH

The total phenol content of sorghums is significantly correlated with antioxidant activity (Awika et al., 2003b; Dicko et al., 2005; Dykes et al., 2005). These findings are in contrast to those of Kamath et al. (2004) who did not find a correlation between phenols and antioxidant activity, simply because they used only white sorghums, which are quite low in phenols. Sorghums with a pigmented testa increase antioxidant activity (Awika et al, 2003b; Dicko et al., 2005; Dykes et al., 2005) due to the presence of tannins, which are more potent antioxidants than monomeric phenolic compounds (Hagerman et al., 1998). Dicko et al. (2005) reported a correlation between antioxidant activity and tannins ($r = 0.79$); Dykes et al. (2005) reported a strong correlation between antioxidant activity and flavan-4-ol levels ($r = 0.88$) among non-tannin sorghums with a red pericarp. Sorghums containing condensed tannins have consistently shown the highest antioxidant activity in vitro. They approach or exceed the antioxidant levels of fruits and vegetables (**Table 10**).

Table 10: Antioxidant activity (ORAC) of sorghum grain and bran compared to common fruits and vegetables

Commodity	ORAC (mmol TE/g, dry wt)
Tannin sorghum (grain)[a]	868
Tannin sorghum (bran)[a]	3124
Black sorghum (grain)	219
Black sorghum (bran)	1008
Red sorghum (grain)	140
Red sorghum (bran)	710
White sorghum (grain)	22
White sorghum (bran)	64

Blueberry, lowbush	842
Strawberry	402
Plum	495
Watermelon	18
Apple, red delicious	295
Orange, navel	137
Broccoli	173
Carrot	108
Onion, red	93
Sweet pepper, green	105
Radishes	217
Potatoes, russet	63

[a]Sumac variety. Awika et al. (2003b); Wu et al. (2004).

Factors affecting the type of phenols and their composition are partially understood in sorghum. Our partial understanding, however, could enable the breeding of nontannin and tannin sorghums with significantly higher levels of antioxidants. Dicko et al. (2005) and Dykes et al. (2005) reported that antioxidant activity was increased when sorghums had purple/red secondary plant colour, a black or dark-red, thick pericarp, and a pigmented testa with a spreader gene. Additionally, decortication of sorghum to produce sorghum bran increases sorghum phenols and antioxidant activity 3–5 times over the original grain (Awika et al., 2003b, 2005). Several health benefits of sorghum have been reported (Awika and Rooney, 2004). For example, Klopfenstein et al. (1981) reported that guinea pigs fed low tannin sorghum had significantly lower cholesterol levels than those fed whole wheat, rolled oats, or pearl millet diets. In addition, Rooney et al. (1992) reported that sorghum and pearl millet brans were excellent bulking agents compared to wheat. They also reported that millet brans have better cholesterol-lowering properties than wheat, white sorghum, or brown sorghum brans. Rats fed sorghum, proso millet or buckwheat at 30% (w/w) of the total diet increased HDL cholesterol without changing total cholesterol level (Cho et al., 2000). Lee and Pan (2003) reported that fish feed diets containing tannin sorghum distillery residues gave better blood-thinning activities in gray mullet than the control diet during the winter season. Tannin sorghums are slowly digested. Some cultures in Africa prefer tannin sorghums since it contributes a longer period of satiety or fullness compared to other cereals, which could be due to its slow digestibility (Awika and Rooney, 2004). This has potential applications in foods for diabetics. Sorghum and millet have anti-carcinogenic properties. For example, Van Rensburg (1981) and Chen et al.

(1993) reported that populations consuming sorghum and millet had lower incidences of esophageal cancer compared to those consuming wheat or maize. However, Morton (1970, 1972) reported that there was an association between high tannin sorghum consumption and human esophageal cancer, but these studies were criticized due to inadequate experimental design (Awika and Rooney, 2004; Yu and Swaminathan, 1987). Grimmer et al. (1992) showed that polymeric tannins from sorghum had higher anti-mutagenic activity than the lower molecular weight tannins. Gomez-Cordoves et al. (2001) showed that sorghum tannins increased melanogenic activity without increasing total melanin and reduced the formation of human melanoma colony cells. A recent study by Turner et al. (2006) showed that black and tannin sorghum bran reduced colon carcinogenesis in rats. In their study, rats fed diets containing black or tannin sorghum bran had fewer aberrant crypts than those fed diets containing cellulose or white sorghum bran. The reduction in colon carcinogenesis could be due to the antioxidant activity of the black and tannin sorghum bran (Turner et al., 2006). Studies are needed to determine which compounds are responsible for the anti-carcinogenic effects of sorghum. The antioxidant activity of millet phenols and their health benefits have also been reported. For instance, in Japanese barnyard millet, the antioxidant activity of luteolin was nearly equal to that of quercetin; however, the activity of tricin was lower than luteolin (Watanabe, 1999). Finger millet is a potent source of antioxidants and has potent radical-scavenging activity that is higher than that of wheat, rice, and other millets; these results corresponded to their phenolic content (Sripriya et al., 1996). The brown or red variety of finger millet, which is commonly available, had higher activity (94%) than the white variety (4%) using the DPPH method (Sripriya et al., 1996). Kodo millet quenched DPPH by nearly 70% higher than other millets (15–53%); white millet varieties had lower activity (Hegde and Chandra, 2005). Luteolin, a flavone present in sorghum and millets, has been reported to have antioxidant, anti-inflammatory, cancer-preventive, anti-arrhythmic properties (Duke, 1992; Watanabe, 1999). Tricin has been reported to have anti-tumor and antileukemic properties (Lee et al., 1981; Watanabe, 1999).

Sorghum contains large quantities of phenolics and other compounds of use in human foods to prevent deterioration of health. Different phenols with varying properties exist, but relatively little effort to demonstrate the potential of these compounds in human health has been made. Of all the cereals, sorghum has the potential to be bred specifically to produce high levels of different phenols that can be easily concentrated by simple processes. These special sorghums have reasonable grain yields and agronomic characteristics that make them productive and economical to produce. The future is bright, but significant investment in evaluation and feeding trials to demonstrate their health promoting properties is required. Millets, with the exception of finger millet, are not as promising. However, they have been the subjects of little if any research.

ANTIOXIDANT PROPERTIES OF SORGHUM PHENOLS AND THEIR BIOAVAILABILITY

Currently antioxidant activity is the most common in vitro parameter used to assess or predict potential benefits of plant phytochemical compounds. However, correlations between in vitro antioxidant activity and actual health benefits are unknown. Such in vitro antioxidant data ignore other potentially beneficial or harmful effects of phytochemicals like modification of enzyme activity and/or cell signaling pathways. For example, vitamin C and E, and the carotenoids, which were previously recognized only for their antioxidant characteristics, were shown to induce other biological responses (Astley, 2003). Antioxidant activity data are also hard to compare since there are no standardized methods; the methods currently used do not always agree in terms of ranking samples for antioxidant efficacy. Additionally, measured antioxidant activity in vitro tells us nothing about release and uptake of the compounds, as well as their distribution and metabolism in the body. However, antioxidant activity data still provide useful information for screening plant materials and products with desirable compounds and properties that can be used for further biological testing. In sorghum, phenol content correlates most strongly with antioxidant activity measured by various methods (**Table 11; Figure 10**) indicating the phenols are largely responsible for the activity (Awika, 2003).

Table 11: Phenol contents and antioxidant activities (lmol TE/g sample, dry wt) of sorghum and sorghum products measured by different methods.

Sorghum	ORAC[a]	ABTS[b]	DPPH[c]	Phenols[d]
White grain	22	6	6	0.8
White bran	64	28	21	4.8
Red grain	140	53	28	6.6
Red bran	710	230	71	19.9
Black grain	220	57	41	6.4
Black bran	1000	250	184	26.0
CSC3*R28 grain	450	108	118	12.3
CSC3*R28 bran	2400	512	495	54.9
Sumac grain (brown)	870	226	202	19.8
Sumac bran (brown)	3100	768	716	66.3
CV%	6.8	3.5	5.3	5.97

[a]Oxygen radical absorbance capacity, fluorescein used as a probe. [b] 2,20-Azinobis (3-ethyl-benzothiaziline-6-sulfonic acid); activity was measured after 30 min reaction in pH 7.4 phosphate buffer saline. [c]2,2-Diphenyl-1-picrylhydrazyl; activity was measured after 8 h reaction in methanol. [d] mg GAE/g (Folin–Ciocalteu method). Samples were extracted in 70% aqueous acetone. Adapted from Awika et al. (2003b).

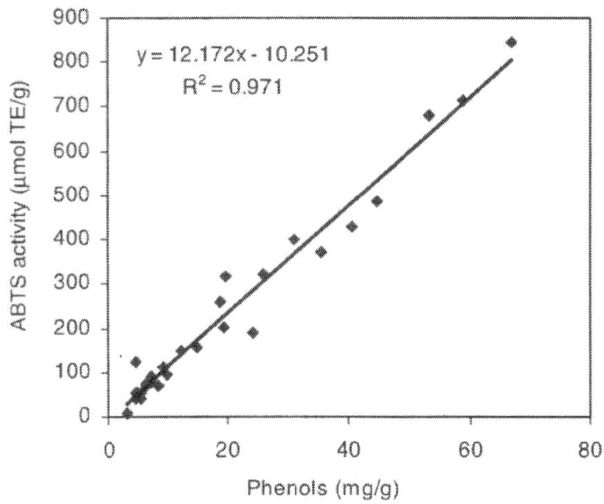

Figure 11: Correlation between antioxidant activity (ABTS) and level of phenols in sorghum and sorghum brans. Samples were extracted in acidified methanol (1% HCl) (Awika et al., 2003b).

CONDENSED TANNINS

Tannins from sorghum show powerful antioxidant activity in vitro (Hagerman et al., 1998; Riedl and Hagerman, 2001). We also found that tannin (brown) sorghums had antioxidant activities higher than most non-tannin sorghums (**Table 11**). High MW tannins have the greatest antioxidant activity in vitro (on a molar basis) among natural antioxidants (Hagerman et al., 1998; Bors et al., 2000). This was attributed to the proximity of many aromatic rings and hydroxyl groups and the fact that tannins were not able to act as prooxidants (Hagerman et al., 1998). Procyanidin o-quinone is capable of producing oligomeric compounds through various coupling reactions that retain the number of hydroxyl groups, unlike the simple flavonoid o-quinones that can act as prooxidants by forming reactive oxygen species through futile redox cycling (Bors et al., 2000). A major concern about tannins though is that they may not be biologically effective antioxidants due to their large molecular size and their tendency to bind food molecules into insoluble complexes. However, Riedl and Hagerman (2001) demonstrated that even when complexed with proteins, sorghum tannins retained at least 50% of their antioxidant activity. Such protein-complexed tannins may serve as free radical sinks in the digestive system thus sparing other antioxidants.

Recent data show that tannins are more bioavailable than previously thought (Ross and Kasum, 2002). Deprez et al. (2001) reported that procyanidins, up to trimers, could be absorbed by the intestinal cell monolayer. Additionally, Spencer et al. (2000) reported that the interflavan bond in the procyanidins was unstable in a simulated gastric juice (pH 2) environment, which degraded the higher MW procyanidins to monomers and dimers. This potentially improves bioavailability of the procyanidins. However, Rios et al. (2002) reported that procyanidins are not degraded (are stable) during gastric transit in humans, hence only the monomers and dimers may be directly bioavailable. Still, part of the non-absorbed tannins is degraded by colon microflora into phenolic acids that are absorbed and may provide additional benefits (Pietta et al., 1997; Deprez et al., 2000; Tapiero et al., 2002).

PHENOLIC ACIDS

Phenolic acids are more readily absorbed than other phenols from food due to their small molecular sizes (Scalbert et al., 2002). These compounds are readily extractable from fruits and vegetables, where they significantly exist in free forms. However, in sorghum and other cereals, most of the PA are esterified to cell wall components and can only be extracted in meaningful quantities by alkaline hydrolysis. Such bound forms of phenolic acids were, until recently, considered unavailable for absorption. However, Kroon et al. (1997) and Andreasen et al. (2001a,b) demonstrated that human colonic esterases (mostly of microbial origin) are capable of releasing esterified diferulates and other hydroxycinnamic acids from cereal brans. This implies that the bound PA are potentially bioavailable and the actual contribution of PA to health benefits associated with consumption of whole grains may be greater than previously assumed. In sorghum, bound PA generally account for over 85% of total PA (Hahn, 1984; Waniska et al., 1989). Adom and Liu (2002) also reported that more than 90% of ferulic acid (the most abundant PA in cereals) in corn, wheat, rice and oats exists in bound form.

ANTHOCYANINS

Anthocyanins were reported to have low absorption compared to other flavanoids (Wu et al., 2002). However significant absorption of these compounds was demonstrated (Lapidot et al., 1999; Matsumoto et al., 2001; Milbury et al., 2002). Most of these data were obtained for fruit anthocyanins which are thought to contribute significantly to the health benefits of fruit consumption. We have not found any work reporting bioavailability of the 3-deoxyanthocyanidins commonly found in sorghum. This is probably because these relatively rare anthocyanins were not considered to be of economic interest. The high antioxidant capacity of black sorghums and their brans were correlated with their anthocyanin contents (**Figure 12**). Hence, anthocyanins may contribute significantly to any potential health benefits of these sorghums. Boveris et al. (2001) demonstrated that a 3-deoxyanthocyanidin (apigeninidin) isolated from soybean had strong dose-dependent quenching ability against ascorbyl and lipid radicals. Anthocyanins from fruits have been shown to possess several therapeutic benefits, including vasoprotective and anti-inflammatory properties (Lietti et al., 1976), anti-cancer and chemoprotective properties (Karaivanova et al., 1990), as well as anti-neoplastic properties (Kamei et al., 1995). The sorghum anthocyanins should be investigated for any unique health properties.

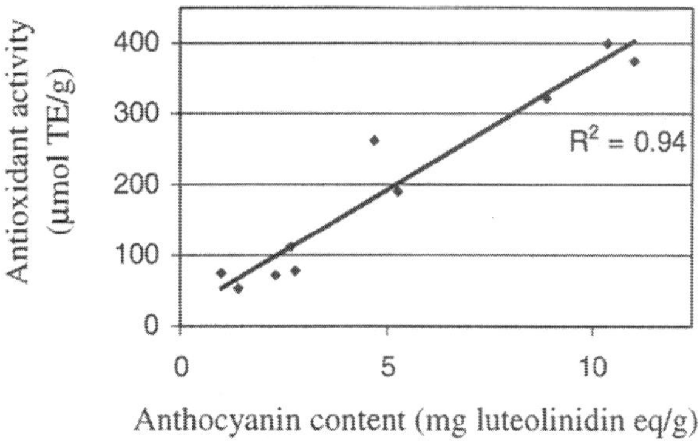

Figure 12: Correlation between anthocyanin content and antioxidant activity (ABTS) of black sorghums and their brans. Anthocyanins were determined by the pH differential method. Acidified methanol (1% HCl) was the extracting solvent.

COMPARING SORGHUM WITH FRUITS

Blueberries are considered an excellent source of antioxidants (Heinonen et al., 1998; Prior et al., 1998) and are commonly used as ingredients in various baked foods (e.g., muffins). Antioxidant activities of blueberries and other common fruits are compared to those of sorghum brans in **Table 12**. The sorghum brans show significantly higher values than the fruits. The high ORAC (oxygen radical absorbance capacity) levels in sorghum brans demonstrate a high potential of the sorghum brans compared to fruits as a source of natural antioxidants. The sorghum fractions can provide high antioxidant properties when used as ingredients in cereal-based foods. However, comparable antioxidant properties do not necessarily translate into comparable health benefits.

Table 12: Antioxidant (ORAC) activity in sorghum brans relative to common fruits

Sample	**ORAC[a] (dry wt)**
Black sorghum bran	1010
Brown sorghum bran	2400 – 3100
Blueberries	87 – 870
Strawberries	356 – 400
Plums	452 – 600
Grapes	100
Watermelon	15
Orange	80 - 150

[a]μmol TE/g, using fluorescein as a probe. [b] Unpublished data, courtesy of Wu, X., Arkansas Children's Nutrition Center, USDA-ARS, AR. Awika (2003); Wu et al. (2002)

PHYTOSTEROLS

Phytosterols are cholesterol (44)-like compounds that are structural components of plant cell membranes (**Figure 13**). In cereals grains they are mostly found in bran and are extractable as part of bran oil waxes. There is a considerable interest in these compounds due to their promotion of cardiovascular health, especially through their cholesterol-lowering properties. Cereal brans reported to have high levels of these compounds include rice (Rogers et al., 1993; Dunford and King, 2000; Fang et al., 2003) and corn (Moreau et al., 1996; Singh et al., 2003). The phytosterols in cereals exist in free forms, as esters of fatty acids or hydroxycinnamic acids (usually ferulate), or conjugated with sugars (mostly glucose). In sorghum the free phytosterols identified include sitosterol (45), campesterol (46) and stigmasterol (47) (Avato et al., 1990). Esterified forms (48–50), with fatty acid chains of C14–C24 (Avato et al., 1990) and ferulates (Singh et al., 2003) were also identified in sorghum. Similar compounds are found in rice and corn brans (Rogers et al., 1993; Moreau et al., 1996). Stanol forms of the phytosterols (without a double bond at position 5) have also been reported in cereals (Ostlund, 2002). These compounds are not as commonly studied as the sterols, but are reported to offer similar health benefits. Ostlund (2002) reported that the stanols comprise about 10% of phytosterols in diet. Quantitative data on sorghum phytosterols are limited. Singh et al. (2003) reported total phytosterol levels of 0.5 mg/g for sorghum grain, compared to 0.9 mg/g for corn (**Table 13**). The fiber fraction isolated from sorghum by a corn wet-milling procedure had 0.7–0.8 mg/g phytosterols. Values of 1.7–5.6 mg/g were reported for rice bran under optimized extraction conditions (Xu and Godgers, 2000). However, Singh et al. (2003) reported that the wet milling procedure they used resulted in a loss of 53–73% of the phytosterols in the isolated sorghum fractions relative to whole grain. Additionally the values they reported were only for the hexane-extractable (non-bound) forms of the compounds. Actual phytosterol values for sorghum may be higher with hydrolytic or other optimized extraction procedures commonly used for cereals (Weihrauch and Gardner, 1978; Jonker et al., 1985; Piironen et al., 2002). Particle size of material (Moreau et al., 1996) as well as type of solvent (Xu and Godgers, 2000) are also known to significantly affect amount of extractable phytosterols. However, indications are that sorghums are a viable source of phytosterols (Table 13), and procedures to extract these compounds especially from sorghum bran should be explored.

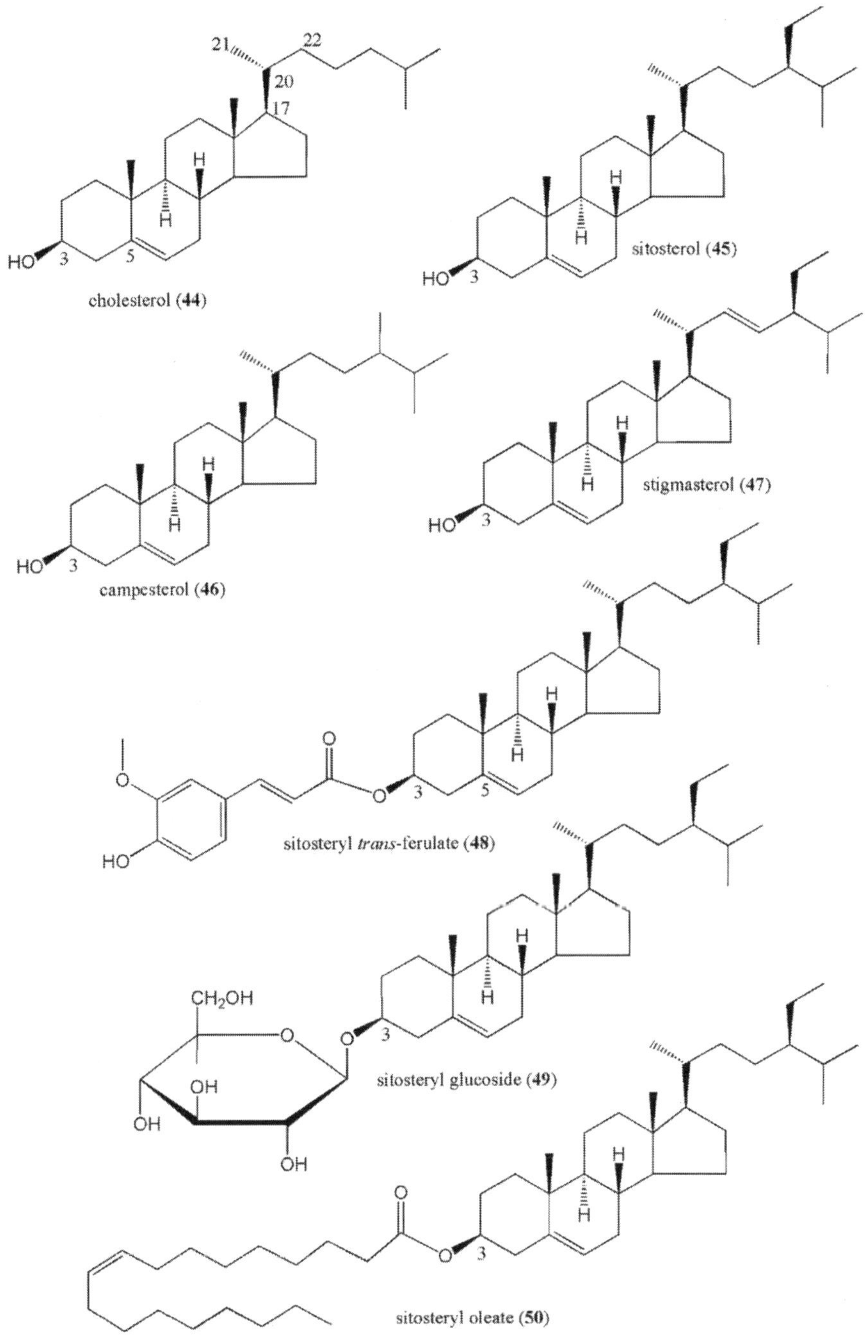

cholesterol (**44**)

sitosterol (**45**)

campesterol (**46**)

stigmasterol (**47**)

sitosteryl *trans*-ferulate (**48**)

sitosteryl glucoside (**49**)

sitosteryl oleate (**50**)

Figure 13: Structure of cholesterol and common phytosterols, and their esters are found in sorghum and other cereals. Notes: Only sitosterol esters are illustrated, but campesterol and stigmasterol esters are also found in sorghum. Stanols are saturated forms of the sterols (single bond at position 5).

Table 13: Phytosterol contents of cereals

Grain	Phytosterols (mg/100 g)
Sorghum	46 - 51[a]
Corn	70 – 88
Barley	55 – 76
Wheat	40 – 74
Oats	35 – 60
Rye	96
Brown rice	72
Buckwheat	96

(b) Singh et al. (2003); (c) Ostlund (2002); (d) Moreau et al. (1996); (e) Piironen et al. (2002); (f) Jonker et al. (1985); (g) Weihrauch and Gardner (1978). a Non-bound phytosterols only (from hexane extracts).

POLICOSANOLS

Policosanols are a mixture of high molecular weight aliphatic alcohols (also called fatty alcohols) that are part of the wax components of plants. The compounds are currently commercially obtained from sugarcane wax by hydrolytic cleavage and further purification (Gouni-Berthold and Berthold, 2002). In sorghum, wax comprises about 0.2% of the grain, generally higher than in other cereals. The policosanols represent 19 - 46% of the sorghum wax, with octacosanol (C28) (1) and triacontanol (C30) (2) as the most abundant (Bunger and Kummerow, 1951; Dalton and Mitchell, 1959; Seitz, 1977; Avato et al., 1990). This translates to approximately 38–92 mg of policosanols in every 100 g sorghum grain. Weller and Hwang (2003) reported that these compounds may be the most commercially valuable component of sorghum grain based on their current market value. The policosanols have not been extensively studied in cereals, and most of the available literature is based on sugarcane wax policosanols.

However, the compounds are gaining rapid popularity due to their cholesterol lowering ability and other sources, including sorghum, will definitely be of considerable interest.

The policosanols have cholesterol-lowering potency comparable to that of statins (currently popular but expensive and potentially harmful drugs) (McCarthy, 2002). Castano et al. (2002) reported that 10 mg/day of policosanol was more effective than 20 mg/day of lovastatin in reducing LDL cholesterol and raising HDL cholesterol levels. Other studies have shown similar benefits (reviewed by Gouni-Berthold and Berthold, 2002; Pepping, 2003). These authors also report that the policosanols present no toxic effects even at high doses. Other positive benefits provided by policosanols include effects on lipid peroxidation, platelet aggregation and smooth muscle cell proliferation (Fraga et al., 1997; Castano et al., 2002; Gouni-Berthold and Berthold, 2002). The policosanols are destined to gain importance as natural, safe and effective dietary alternatives to statin medication. Efficacy and economic potential of the sorghum policosanols should be investigated.

SORGHUM PHYTOCHEMICALS AND HUMAN HEALTH

Sorghum and cardiovascular disease

Cardiovascular disease (CVD) is the number one killer in the USA (Sistino, 2003). Various epidemiological data indicate that whole grain consumption significantly lowers mortality from CVD (Kushi et al., 1999; Slavin et al., 2000; Anderson, 2003). The phytosterols in the cereal brans are believed to contribute to beneficial effects. Other components of the whole grains, including polyphenols and fiber, also play a role in CVD prevention. For example, a cholesterol-lowering effect of tea and grape catechins and tannins is widely reported (Lin et al., 1986; Tebib et al., 1997; Santos-Buelga and Scalbert, 2000). We did not find any study that directly assesses effects of sorghum polyphenols on cholesterol.

In vivo studies on effects of sorghum on CVD are scarce. Klopfenstein et al. (1981) reported a cholesterol lowering effect of low-tannin sorghum grain when fed to guinea pigs at 58% of diet. This effect was greater than that produced by wheat, rolled oats or pearl millet. More recently, Cho et al. (2000) found that sorghum and proso millet hexane extracts inhibit rat liver microsomal 3-hydroxy-3-methylglutaryl CoA (HMGCoA) reductase in a dose-dependent manner. They also observed that fecal bile acid excretion, as well as levels of HDL cholesterol ere increased, without a change in total cholesterol, when whole sorghum, proso millet or buckwheat were fed to rats at 30% of diet. Rooney et al. (1992) on the other hand observed that when sorghum bran was added to rat diet to achieve a 6% actual sorghum fiber, both high-tannin and non-tannin sorghums as well as wheat bran increased blood serum total cholesterol in rats. However, they did not measure changes in HDL and LDL cholesterol components. It is apparent that more data is necessary on how the diverse sorghum types affect cholesterol metabolism in controlled animal studies. Available data showing sorghum are a significant source of phytosterols and policosanols (Dalton and Mitchell, 1959; Seitz, 1977; Avato et al., 1990; Singh et al., 2003) will hopefully stimulate additional work in this field.

Benefits of sorghum to cardiovascular health may not be limited to positive effects on cholesterol. Lee and Pan (2003) demonstrated that dietary tannin-sorghum distillery residues inhibited 63–97% of hemoglobin-catalyzed oxidation of linoleic acid in cultured mullet fish compared to soybean (13%) and rice bran (78%). The authors also found that the sorghum residues significantly improved blood-thinning and erythrocyte membrane integrity of the fish blood cells during winter, thus maintaining normal blood fluidity and preventing RBC hemolysis induced by H2O2. They attributed the prevention of RBC hemolysis to the antioxidant activity of the tannins and other polyphenols present in the sorghum residue. Such protective effect on RBC was previously demonstrated for tea (Grinberg et al., 1997; Yang and Koo, 2000) and red wine (Pace-Asciak et al., 1996; Tedesco et al., 2000) polyphenols. The bloodthinning effect is also in agreement with epidemiological data that suggest that polyphenols from tea may reduce risk of stroke (Sato et al., 1989; Keli et al., 1996; Higdon and Frei, 2003).

In general several epidemiological studies show that red wine and tea consumption correlate with reduced risk of CVD (StLeger et al., 1979; Renaud and De Lorgeril, 1992; Higdon and Frei, 2003). The effects are largely attributed to the tannins and other polyphenols in these food products. There is no epidemiological data on tannin sorghum consumption and CVH in humans. This may partly be due to the fact that in the regions with high human sorghum consumption, CVD is not regarded as a major problem since there are other diseases that are of bigger concern. Such a study is important to help validate some of the evidence reported for in vitro and controlled animal studies, and determine the role sorghum may play in fighting CVD. The presence of unique polyphenols in sorghum (discussed earlier) is likely to produce positive benefits.

Sorghum and obesity

Obesity is an ever-increasing problem in the western world and is related to several disease conditions including CVD and diabetes. As a result a plethora of weight loss regimens, diets, pills, etc., has emerged. However, such regimens do not seem to produce the desired effects and obesity cases are on the rise (Hill et al., 2003; Wyatt, 2003). For example, in 1994 approximately 56% of US population was overweight and by 2000 the figure was 65% (Hill et al., 2003). Food (high calorie) consumption and lifestyle (lack of physical activity) account for the bulk of obesity cases. However, a radical change in either of these factors, as demanded by many weight loss strategies, is not easy to sustain for long periods of time, hence the failure of the strategies to produce an impact on obesity in the population.

Numerous reports on reduced weight gain of animals (rats, pigs, rabbits, poultry) fed high tannin sorghum are available (Jambubathan and Mertz, 1973; Featherson and Rogler, 1975; Cousins et al., 1981; Lizardo et al., 1995; Al-Mamary et al., 2001; Muriu et al., 2002). The mechanisms by which tannin sorghums reduce nutritive value include binding of food proteins (Haslam, 1974; Hagerman and Butler, 1981) and carbohydrates (Naczk and Shahidi, 1997) into insoluble complexes that cannot be broken down by digestive enzymes. Another mechanism involves the direct binding of digestive enzymes including sucrase, amylases, trypsin, chymotrypsin and lipases (Lizardo et al., 1995; Carmona et al., 1996; Nguz et al., 1998; Al-Mamary et al., 2001), thus inhibiting their activity. Inhibition of intestinal brush-border bound amino acid transporters by sorghum tannins was also reported (King et al., 2000). There is evidence that the higher

DP tannins are more involved in these interactions that the low DP ones (Bacon and Rhodes, 1998; Sarni-Manchado et al., 1999).

Effects of the sorghum tannins on animal weight gain depend on levels fed as well as animal species. Al-Mamary et al. (2001) found addition of 1.4% catechin equivalents (CE) sorghum to rabbit diet had no effect on growth rate and weight gain, whereas at a CE of 3.5%, there was a marked decrease in live weight gain and feed conversion ratio. Cousins et al. (1981) reported a 10% reduction in feed efficiency (relative to corn) when high tannin (3.1–3.4% CE) sorghums were fed to pigs. Low tannin sorghum (0.8% CE) had similar feed efficiency as corn.

Lizardo et al. (1995) also observed a 10% reduction in weight gain when pigs were fed tannin-sorghum diet compared to corn. Broiler chicks fed low tannin sorghums (0.6–0.9% CE) had similar weight gain and feed efficiency compared to corn (Ambula et al., 2001). However, these authors observed an average of 50% reduction in weight gain when high tannin sorghums (2.7–3.5% CE) were fed to the broiler chicks. Such data suggest that there is a threshold for the inhibitory effects of sorghum tannins, and that animal species are affected differently.

Despite the evidence on animal studies, there is no reported work on how these attributes of tannin sorghums can be used to help lower calorie intake in overweight humans. Epidemiological data on humans in this area is also lacking, probably because in the places where tannin sorghums are consumed, obesity is often not a problem due to different lifestyles, daily food (calorie) intake and other factors. In certain cultures in Africa tannin sorghum is preferred to other cereals since it has a longer "staying power" in the stomach, i.e., offers durable satiety value. This property may be related to the slow digestibility and nutrient release from the tannin-complexed food matrix and should be investigated.

It is essential to assess the possibility of using this information to fight obesity in humans. Grain-based foods (breakfast cereals, bread, cookies, extruded snacks, etc.) are common and widely consumed in many parts of the world. High tannin sorghum grains and their milled fractions (bran) can find use as parts of ingredients in such foods. Brans from high tannin sorghums were incorporated in bread and cookies at up to 15% and 30%, respectively, without significant differences in texture or flavor profiles compared to whole wheat products (Gordon, 2001; Mitre-Dieste et al., 2000). Cereal-based foods are a viable means of fighting obesity since they are consumed more consistently than most other foods and contribute a significant portion of the daily calorie intake. However, to use tannin sorghums in human diet to help fight obesity, the following need to be determined:

1. Whether the effects observed in animals are reproducible in humans.
2. The levels of tannins in sorghum necessary to produce desired effects.
3. Potential side effects of using tannin sorghums and their fractions at such levels, e.g., effects on availability of other essential micronutr-

ients, especially divalent minerals like iron (which is chelated by tannins), and how the effects can be overcome.

4. How various food processing conditions affect potential activity of the tannins?

Sorghum and cancer

Positive effects of sorghum and/or millet consumption on cancer have been documented. Van Rensburg (1981) reported that sorghum consumption consistently correlated with low incidences of esophageal cancer in various parts of the world (including several parts of Africa, Russia, India, China, Iran, etc.) whereas wheat and corn consumption correlated with elevated incidences. Such regions also had deficiencies of certain minerals and vitamins in their diets. In attempting to explain this phenomenon, the author proposed (with considerable evidence) that the nutrient deficiencies were responsible for the high esophageal cancer incidences, and that sorghum and millet consumption promoted resistance to esophageal cancer risk. Chen et al. (1993) reported similar results from epidemiological data from Sachxi Province, China. These authors studied 21 communities within the province over a period of 6 years and found that regions that consumed highest amounts of sorghum, and to a lesser extent millet, had 1.4–3.2 times lower mortality from esophageal cancer than areas that primarily consumed wheat flour or corn. Consumption of other foods like alcohol, tea, meats and vegetables did not contribute significantly to esophageal cancer mortality. These evidences suggest presence of anti-carcinogenic compounds in sorghum that are either lacking or are not present in significant quantities in wheat or corn.

On the other hand, consumption of plants containing tannins (tea, sorghum, betel nuts, etc.) was previously implicated in incidences of cancer in the upper digestive tract (Morton, 1970, 1972). Morton (1970) implicated high incidences of esophageal cancer in certain parts of South Africa, China and Russia on consumption of high tannin sorghums. Oterdoom (1985) theorized that the carcinogenic effect of the high tannin sorghums reported by Morton (1970) was due to the fact that "tannins destroyed proteins both in the mucosa and in enzymes". However, the carcinogenic evidence suggested by Morton (1970) was criticized by Yu and Swaminathan (1987), and we concur, for lack of acceptable experimental design or control for confounding variables. Actually, Morton (1970) did not present any data to back her claim, but merely theorized that the common denominator in regions with elevated risk of esophageal cancer was probably the consumption of plants containing tannins.

In vitro studies have also revealed anti-carcinogenic properties of sorghum. Grimmer et al. (1992) demonstrated anti-mutagenicity of sorghum polyphenol extracts. They found the high MW procyanidins (tannins) had the highest anti-mutagenic activity compared to lower MW tannins. Gomez-Cordovez et al. (2001) showed that sorghum tannins had anti-carcinogenic activity against human melanoma cells, as well as positive melanogenic activity (melanogenesis is believed to help protect human skin against UV irradiation damage (Eller et al., 1996)). The authors did not observe such melanogenic activity in red wine extracts. On the other hand, Parbhoo et al. (1995) reported that sorghum procyanidin extracts may induce cytochrome P-450, a protein that is capable of converting certain promutagens to mutagenic derivatives, in rat liver.

There is plenty of literature available (and often conflicting) on properties of various polyphenol-rich foods, especially tea and red wine/grapes in relation to various types of cancer in humans (reviewed by Chung et al., 1998; Santos-Buelga and Scalbert, 2000; Higdon and Frei, 2003). The overall trend suggests positive (anti-carcinogenic) effects of the polyphenol-rich products in diet rather than negative (carcinogenic) ones, but a consensus is far from deducible. For example, even though tea is reported as anti-carcinogenic in many in vitro and epidemiological studies, Higdon and Frei (2003) in an excellent review concluded that when all is considered, available data do not support the notion that tea is protective against cancer. Yang et al. (2001) also provided useful insight on the limitations of various cancer-related studies.

For sorghum, available data on cancer are too limited to draw reasonable conclusions. However, the corroborative epidemiological evidences reported by Van Rensburg (1981) and Chen et al. (1993) against esophageal cancer warrant some follow up. Additional *in vitro* data as well as controlled animal studies are necessary to understand how the levels and composition of polyphenols in sorghum affect cancer, and which specific components are responsible. Whole grain consumption has long been correlated with reduced risks to other forms of digestive tract cancer, especially colon cancer. How much of these effects are contributed by the dietary fiber or the phytochemicals concentrated in the brans of the grains is still unknown.

Perspective

Sorghum has a diversity of phytochemicals with a potential to significantly impact human health. The sorghum phytochemicals show high antioxidant activity against different free radicals in vitro relative to fruits and vegetables, and may offer similar benefits attributed to fruits and vegetables. Information on how sorghum phytochemicals affect human health is scarce. However, overall epidemiological evidence suggests sorghum has anti-carcinogenic properties when consumed regularly in diet. Animal studies also indicate that sorghum consumption promotes cardiovascular health better than other cereals. It is absolutely essential to determine if the positive effects observed in animals can be reproduced in humans; given cardiovascular disease is a leading killer in the west.

The most negatively reported effect of sorghum tannins in animals is reduced weight gain. However, with obesity a major and an ever-increasing problem in the developed world (more than 60% of Americans are reported to be overweight), this attribute of sorghum tannins has the potential of helping alleviate the problem. But we first need to determine if and what levels of tannin sorghum fractions in food can produce a desired weight reduction in humans and address any potential side effects that such levels may create.

Overall, the most effective (cost or otherwise) solutions to many major human health conditions lie in the natural components of foods we eat, rather than expensive medical intervention. The challenge is to find and incorporate a balance of the functional ingredients in everyday foods at adequate levels. Sorghum, which is currently underutilized, is definitely worth attention as a source of health-promoting phytochemicals for such foods. Some potential applications of sorghum include:

1. Direct food use. Whole sorghum grain can be used in baked, extruded and other cereal-based products (bread, cookies, expanded snacks, pasta, breakfast cereals, etc.) as partial or complete substitutes for other cereals. The sorghum brans can be used to fortify bread, cookies and other snacks, to improve the phytonutrient content, as well as dietary fiber and sensory properties. We previously demonstrated that pigmented sorghum brans in combination with other ingredients produce acceptable quality, dark-coloured baked goods that appeal to consumers due to their natural 'healthy' appearance (no need for artificial darkening, e.g., as caramel is used to darken rye bread).
2. Extraction of active components for commercial use. Most phytochemicals in sorghum are concentrated in bran fractions. These fractions are easily separated from sorghum by decortication and can then be used to extract the various phytochemicals for dietary supplementation, food quality improvement or therapeutic applications. For example, sorghum anthocyanins, which were reported as more stable than the fruit anthocyanins, can be used as natural food colours with functional properties. The phytosterols and policosanons are found in sorghum bran and spent distiller's grain in relatively large quantities; these sorghum fractions may provide a low cost source of these valuable compounds. Sorghum tannins can be extracted for use as antioxidant supplements as well as anti-caloric agents for obese individuals. Other phenolic antioxidants from sorghum can also be used as natural food preservatives, antioxidant supplements and therapeutic agents, among other uses.

4. Diabetes Control

Excessive carbohydrates break down into simple sugars and wreak havoc on the glucose levels in the body, leading to diabetes or causing chaos for those people who already suffer from this disease. However, the tannin-rich bran of sorghum actually has enzymes that inhibit the absorption of starch by the body, which can help to regulate insulin and glucose levels in the body. By keeping these levels balanced, diabetics won't suffer as many plunges and spikes in their glucose levels, thereby preventing diabetic shock and other health complications.

Chemical and phenolic components in three Korean sorghum varieties (**Figure 14**) which are playing major role for controlling diabetes in rat experiments. Major chemical components differed among the three sorghum varieties (**Table 14**). Starch content of the tested samples was similar to that of most sorghum cultivars (~70%), except Chal Sorghum had only 66.3% starch. Chal Sorghum had significantly higher (N13%) crude protein content than the rest of the samples. Twenty-four of the 29 phenolic components monitored were detected; only major peak B and C were not detected in the sorghum studied (**Figure 15**). Overall, phenolic components concentration was greater in HGS than CS and HS. Sorghum varieties also contained different types of phenolic components in varying numbers, ranging from A to E. Two peaks were more detected in only HGS (e.g., B and C). The above indicates that two peaks (B and C) may play a more important role in the antidiabetic effects of Korean sorghum. So, single component evaluation and further investigations on the effects on antidiabetic are necessary.

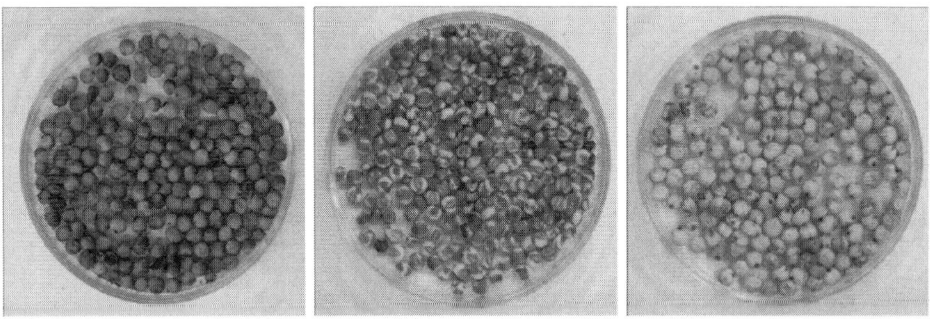

HGS **CS** **HS**

Figure 14: Photograph of three Korean sorghum cultivars. Hwanggeumchal sorghum (HGS), Chal sorghum (CS), Heuin sorghum (HS).

Table 14: Chemical composition of Korean sorghum varieties (%, db)[a].

Korean sorghum	Starch	Crude protein	Crude fat	Crude fiber	Ash
Hwanggeumchal sorghum	73.2[a]	11.63[b]	2.23[c]	2.62[b]	1.43[c]
Chal sorghum	66.3[b]	13.42[a]	3.78[a]	2.82[a]	1.92[a]
Heuin sorghum	72.8[a]	11.23[b]	3.41[b]	1.83[c]	1.61[b]

[a]Means in a column with different letters differ (Pb0.05).

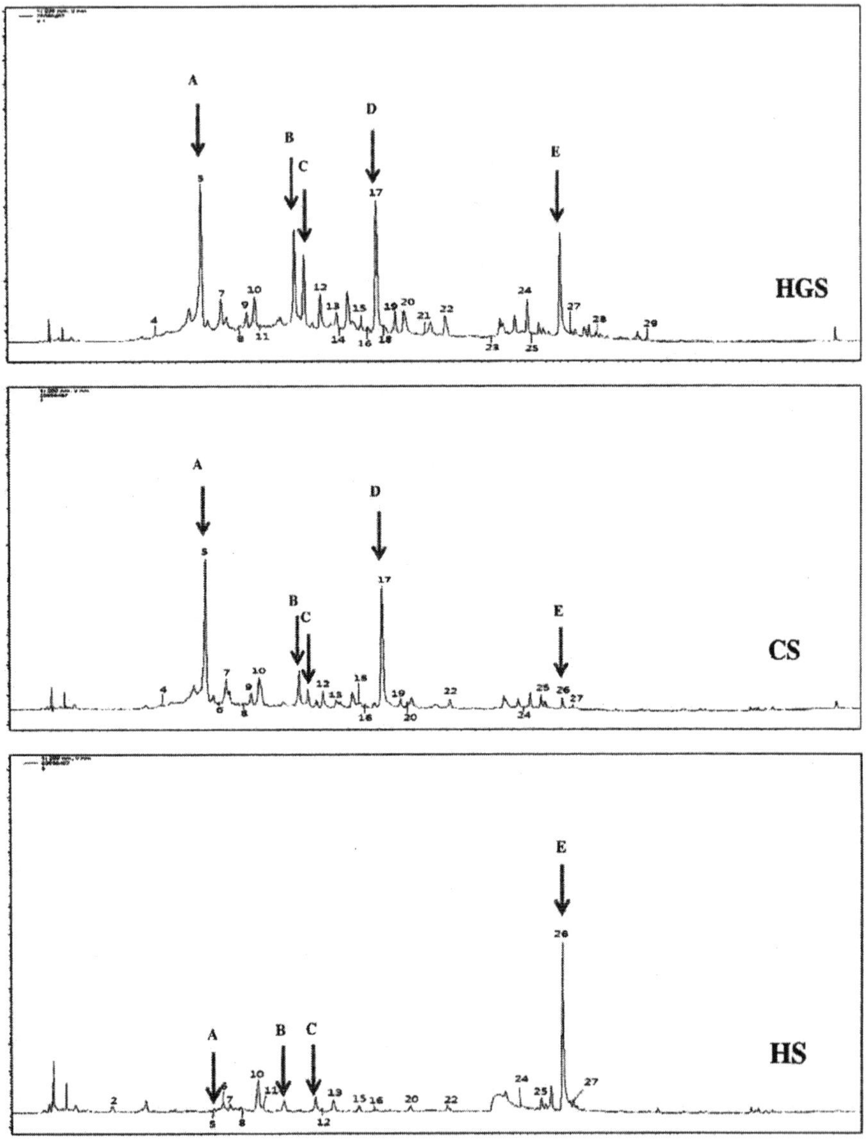

Figure 15: Chromatogram of phenolic component analysis by HPLC. Hwanggeumchal sorghum (HGS), Chal sorghum (CS), Heuin sorghum (HS). 1. Gallic acid; 2. Pyrogallol; 3. Homogentisic acid; 4. Protocatechuic acid; 5. Gentisic acid; 6. Chlorogenic acid; 7. p-Hydroxybenzoic acid; 8. Vanillic acid; 9. Caffeic acid; 10. Syringic acid; 11. Vanillin; 12. p-Coumaric acid; 13. Ferulic acid; 14. Veratric acid; 15. m-Coumaric acid; 16. Benzoic acid; 17. O-oumaric acid; 18. Resveratrol; 19. t-Cinnamic acid; 20. β-Resorcylic acid; 21. Rutin; 22. Hesperidin; 23. MyricetinG; 24. Quercetin; 25. Naringenin; 26. Hesperetin; 27. Formononetin; 28. Biochanin A; 29. Naringin.

ANTIDIABETIC ACTIVITIES OF THREE KOREAN SORGHUM VARIETIES

Loss of body weight is a characteristic condition in diabetes, owing to defect in glucose metabolism and excessive breakdown of tissue protein (Sezik, Asla, Yesilada and Ito, 2005). After 14 days of treatment, group 1~7 increased significantly ($p<0.05$) in weight, when compared to day 0. The STZ-induced diabetic rats had significantly lost weight, compared to the control. However, diabetic rats that were subsequently treated with sorghum phenolic extracts did not significantly lose weight compared to the control (**Table 15**).

Table 15: Effect of animal body weight after 14 days of treatment (100, 250 mg/kg b.w./day, i.p.).

Experimental groups	Body weight (g)	
	Day 0	Day 14
1 - Normoglycemic control	211.2±13.2	253.2±11.3*
2 - Control+HGS (100 mg/kg b.w.)	214.0±16.2	245.6±13.0
3 - Control+HGS (250 mg/kg b.w.)	212.1±13.4	234.5±16.3
4 - Control+CS (100 mg/kg b.w.)	228.9±13.5	239.1±13.4
5 - Control+CS (250 mg/kg b.w.)	238.3±16.3	247.3±21.3
6 - Control+HS (100 mg/kg b.w.)	231.4±17.2	248.5±23.7
7 - Control+HS (250 mg/kg b.w.)	231.4±12.7	240.9±23.7
8 - Diabetic	233.2±16.3	214.5±27.8
9 - Diabetic+HGS (100 mg/kg b.w.)	232.4±13.4	223.6±26.8
10 - Diabetic+HGS (250 mg/kg b.w.)	235.7±18.3	221.1±13.6
11 - Diabetic+CS(100 mg/kg b.w.)	229.0±13.5	211.1±21.3
12 - Diabetic+CS (250 mg/kg b.w.)	231.3±19.6	216.4±18.0
13 - Diabetic+HS (100 mg/kg b.w.)	231.0±15.7	219.0±17.9
14 - Diabetic+HS(250 mg/kg b.w.)	222.6±18.1	213.3±19.4
15 - Diabetic+glibenclamide	239.2±11.9	226.7±26.9

Values are mean ± SEM for nine rats. $^{*}p<0.05$ when compared to day 0.

There was a significant elevation in serum glucose, total cholesterol, triglycerides, urea, uric acid, creatinine, AST and ALT while the serum insulin level was decreased significantly in the diabetic rats. **Table 16** shows the effects of the sorghum phenolic extracts on serum glucose, insulin, triglycerides and total cholesterol in normal and diabetic rats. The results showed that serum glucose, triglycerides and total cholesterol of diabetic rats increased while serum insulin decreased, when compared with normal rats.

Table 16: Effect of Korean sorghum phenolic extracts on mean values of serum glucose, insulin, triglycerides and total cholesterol after 14 days of treatment (100, 250 mg/kg b.w./day, i.p.).

Experimental groups	**Glucose (mg/dL)**	**Insulin (IU/L)**	**Triglycerides (mg/dL)**	**Total cholesterol (mg/dL)**
1 - Normoglycemic control	93.3±4.74	11.3±0.34	73.4±4.45	68.8±1.37
2 - Control+HGS (100 mg/kg b.w.)	82.4±2.01	11.9±0.31	111.9±3.42	58.6±2.00
3 - Control+HGS (250 mg/kg b.w.)	78.4±1.41	11.7±0.36	101.4±3.51	57.4±1.96
4 - Control+CS (100 mg/kg b.w.)	92.7±2.46	11.4±0.32	79.7±2.45	68.9±1.43
5 - Control+CS (250 mg/kg b.w.)	92.3±2.44	11.9±0.23	78.7±2.77	66.4±1.21
6 - Control+HS (100 mg/kg b.w.)	92.7±2.53	11.8±0.13	77.2±2.56	63.6±1.75
7 - Control+HS (250 mg/kg b.w.)	91.4±2.89	11.9±0.85	75.8±3.85	62.5±1.67
8 - Diabetic	543.2±6.69	1.02±0.13	114.2±5.34	106.7±2.44
9 - Diabetic+HGS (100 mg/kg b.w.)	538.3±7.84	1.43±0.12	105.3±3.53	93.2±1.45
10 - Diabetic+HGS (250 mg/kg b.w.)	424.2±5.12[c]	1.78±0.16[b]	79.2±4.31[c]	78.4±2.42[c]
11 - Diabetic+CS (100 mg/kg b.w.)	541.2±6.57	1.03±0.17	111.3±4.34	105.3±3.21
12 - Diabetic+CS (250 mg/kg b.w.)	532.1±5.41	1.08±0.11	107.4±4.28	101.9±2.11
13 - Diabetic+HS (100 mg/kg b.w.)	536.1±5.97	1.06±0.35	113.4±2.89	103.2±2.38
14 - Diabetic+HS (250 mg/kg b.w.)	511.2±4.96[a]	1.17±0.56	106.4±4.63	101.7±2.29
15 - Diabetic+glibenclamide	414.3±35.3[b]	1.89±0.07[a]	89.4±3.22[a]	84.2±3.19[b]

Values are mean±SEM for nine rats. [a]$p<0.05$ vs. diabetic group. [b]$p<0.01$ vs. diabetic group. [c]$p<0.001$ vs. diabetic group.

The administration of the HGS extracts at doses of 250 mg/kg body wt. and glibenclamide tended to bring serum glucose (pb0.001), insulin (pb0.001), triglycerides (pb0.001) and total cholesterol (pb0.001) significantly toward normal values, while normal rats did not exhibit any significant alterations in these parameters during the experiment. The HGS phenolic extract treatment groups were found to be similar effective with glibenclamide. The administration of the HGS phenolic extracts at dose of 1 mg/kg body wt did not change serum glucose (pN0.05), insulin (pN0.05), triglycerides (pN0.05) and total cholesterol (pN0.05) levels in normal rats. **Table 17** showed the effect of the sorghum phenolic extracts on the serum urea, uric acid, creatinine, AST and ALT in normal and diabetic rats. The results showed that serum urea, uric acid, creatinine, AST and ALT increased when compared with normal rats. The administration of the HGS phenolic extracts (250 mg/kg body wt.) and glibenclamide significantly decreased the serum urea (pb0.001), uric acid (pb0.001), creatinine (pb0.001), AST (pb0.001) and ALT (pb0.001) when compared with the control diabetic rats. The HGS phenolic extract treatment at a dose of 250 mg/kg body weight was found to be similarly effective with that observed with glibenclamide. The administration of the HGS phenolic extracts (100 mg/kg body wt) did not change the serum urea (pN0.05), uric acid (pN0.05), creatinine (pN0.05), AST (pN0.05), or ALT (pN0.05) levels in normal rats. The results indicated that the HGS phenolic extract treatment significantly decreased serum glucose, triglycerides, cholesterol, urea, uric acid, creatinine, AST and ALT while it increased the serum insulin levels in treated diabetic rats compared with the control diabetic rats. The result also showed that the HGS phenolic extract treatment exhibited a significant decrease in the level of serum lipids in diabetic rats. The data showed that the uric acid levels were increased in diabetic rats. This may be due to a metabolic disturbance in diabetes reflected in the high activities of xanthine oxidase, lipid peroxidation and increased triglycerides and cholesterol (Madinov, Balabolkin, Markov and Markova, 2000). Moreover, protein glycation in diabetes may lead to muscle wasting and an increased release of purine, the main source of uric acid as well as in the activity of xanthine oxidase (Anwar and Meki, 2003). The data showed that the HGS phenolic extract treatment decreased the serum urea and creatinine levels in diabetic rats. Elevation of the serum urea and creatinine, as significant markers, is related to renal dysfunction in diabetic hyperglycemia (Almadal and Vilstrup, 1988). Serum enzymes including AST and ALT are used in the evaluation of hepatic disorders. An increasc in these enzyme activities reflects active liver damage. Inflammatory hepatocellular disorders result in extremely elevated transaminase levels (Hultcrantz, Glaumann, Lindberg and Nilsson, 1986). In accordance with these findings, streptozotocin treatment has a significant role in the alteration of liver functions since the activity of AST and ALT was significantly higher than those of normal values.

As a result, it may be concluded that the HGS phenolic extracts were more useful effective in comparison with glibenclamide in attenuating the increased serum parameters resulting from damage of STZ-induced diabetic rats and that the HGS phenolic extract treatment may be of use as an antidiabetic supplements. Overall, our report from the present analysis should be ground data for undertaking further study, and it is useful information for investigating new sorghum materials for food additives and human health.

Table 17: Effect of Korean sorghum extracts on mean values of serum urea, uric acid, creatinine, AST and ALT for all groups after 14 days of treatment (100, 250 mg/kg b.w./day, i.p.).

Experimental groups	Urea (mg/dL)	Uric acid (mg/dL)	Creatinine (mg/dL)	AST (IU/L)	ALT (IU/L)
1 - Normoglycemic control	38.5±2.1	0.96±0.04	0.47±0.06	126.4±9.3	83.4±4.2
2 - Control+HGS (100 mg/kg b.w.)	39.7±1.4	0.87±0.05	0.61±0.08	156.3±11.9	88.1±3.6
3 - Control+HGS (250 mg/kg b.w.)	44.3±1.1	0.84±0.05	0.74±0.02	138.9±16.8	63.5±4.1
4 - Control+CS (100 mg/kg b.w.)	40.2±1.6	0.98±0.06	0.55±0.03	127.4±11.5	84.1±4.1
5 - Control+CS (250 mg/kg b.w.)	41.3±3.1	0.91±0.13	0.63±0.07	119.9±14.3	79.3±5.2
6 - Control+HS (100 mg/kg b.w.)	41.4±3.1	0.95±0.19	0.57±0.13	126.7±12.5	85.7±3.8
7 - Control+HS (250 mg/kg b.w.)	41.1±4.3	0.91±0.23	0.61±0.21	121.3±11.4	83.5±2.5
8 - Diabetic	117.4±4.8	1.42±0.23	0.64±0.03	211.3±14.3	153.2±13.2
9 - Diabetic+HGS (100 mg/kg b.w.)	110.1±5.8	1.24±0.35	0.59±0.10	196.9±11.6	158.0±13.1
10 - Diabetic+HGS (250 mg/kg b.w.)	89.3±4.1[b]	0.90±0.11[a]	0.54±0.06[a]	141.9±9.8[c]	94.9±7.0[b]
11 - Diabetic+CS (100 mg/kg b.w.)	111.6±3.7	1.39±0.63	0.61±0.11	204.5±11.9	151.6±17.3
12 - Diabetic+CS (250 mg/kg b.w.)	104.2±2.5	1.31±0.32	0.58±0.09	194.2±11.6	143.1±11.4
13 - Diabetic+HS (100 mg/kg b.w.)	108.2±3.6	1.35±0.19	0.59±0.09	201.3±11.4	151.4±11.4
14 - Diabetic+HS (250 mg/kg b.w.)	99.3±4.1[a]	1.21±0.21[a]	0.56±0.02	183.2±12.8 [a]	137.1±9.0
15 - Diabetic+glibenclamide	98.7±5.2	1.13±0.06	0.55±0.02	152.3±15.4	82.5±6.1[b]

Values are mean ± SEM for nine rats. [a]$p<0.05$ vs. diabetic group. [b]$p<0.01$ vs. diabetic group. [c] <0.001 vs. diabetic group.

The study showed the antidiabetic effects from phenolic extracts of three varieties Hwanggeumchal sorghum, Chal sorghum, and Heuin sorghum) from Korean sorghum (*Sorghum bicolour* L. Monech) in normal and streptozotocin-induced diabetic rats. Hwanggeumchal sorghum phenolic extracts in the streptozotocininduced diabetic rats showed significant hypoglycemic activity for 14 days and significantly decreased the serum glucose, total cholesterol, triglycerides, urea, uric acid, creatinine, aspartate amino transferase and alanine amino transferase while it increased the serum insulin in diabetic rats but not in normal rats ($p<0.05$) (at doses of 100 and 250 mg/kg for 14 days). A comparison was made between the action of Hwanggeumchal sorghum phenolic extracts and glibenclamide (600 μg/kg), a known antidiabetic drug. Twenty-four of the 29 phenolic components monitored were detected by high performance liquid chromatography. The antidiabetic effect of the Hwanggeumchal sorghum phenolic extracts was similarly effective with that observed with glibenclamide. Phenolic components were analyzed using high performance liquid chromatography, and a total of 29 phenolic components were detected in the Korean sorghum studied. The relationships of antidiabetic effects and phenolic components of Hwanggeumchal sorghum phenolic extracts were significantly higher than that of Chal sorghum and Heuin sorghum phenolic extracts (Ill-Min Chung et al., 2011).

Effect of sorghum extract on body composition and food intake

Dietary intake data, initial body weight, final body weight, and weights of the liver, epididymal, perirenal, retroperitoneal, and total fat are presented in **Table 2.** Dietary intake decreased significantly in a dose-dependent manner in mice fed 0.5% SE and 1% SE compared to that in the mice fed the LF and HF diets. Final body weight and epididymal, retroperitoneal, and total fat weights were significantly lower in mice fed LF than those in mice fed the HF diet and in those with and without SE. Perirenal fat weight was significantly lower in mice fed the LF diet, 0.5% SE, and 1% SE than that in mice fed the HF diet, suggesting that SE may significantly reduce the accumulation of perirenal fat; however, initial body and liver weights did not significantly differ among the groups.

Table 18: Effect of sorghum extracts on hyperglycemia and dyslipidemia

	LF	HF	0.5% SE	1% SE
Dietary intake (g/day)	2.55 ± 0.04^{a}	2.54 ± 0.03^{a}	2.39 ± 0.05^{b}	2.21 ± 0.04^{c}
Initial body weight (g)	21.35 ± 0.40	21.35 ± 0.36	21.20 ± 0.35	21.35 ± 0.37
Final body weight (g)	26.15 ± 0.61^{a}	41.44 ± 1.78^{b}	39.75 ± 1.75^{b}	38.44 ± 1.81^{b}
Liver (g)	0.90 ± 0.01	1.14 ± 1.78	1.10 ± 0.12	0.95 ± 0.08
Epididymal fat (g)	0.67 ± 0.08^{a}	2.26 ± 0.16^{b}	2.52 ± 0.22^{b}	2.11 ± 0.14^{b}
Perirenal fat (g)	0.03 ± 0.01^{a}	0.07 ± 0.01^{b}	0.04 ± 0.22^{b}	0.04 ± 0.01^{a}
Retroperitoneal fat (g)	0.33 ± 0.04^{a}	1.11 ± 0.10^{b}	1.13 ± 0.13^{b}	1.15 ± 0.14^{b}
Total fat (g)	1.03 ± 0.09^{a}	3.44 ± 0.23^{b}	3.69 ± 0.32^{b}	3.30 ± 0.23^{b}

Table 19 shows serum lipid profiles, liver function, levels of glucose and insulin, and the AUC for the glucose tolerance test in each group. Total cholesterol levels were significantly lower in mice given 0.5% SE and 1% SE than those in HF diet fed mice and was lower in mice given 1% SE than those in mice fed the LF diet. LDL-cholesterol level was significantly ($P < 0.001$) lower in mice given 0.5% SE and 1% SE than that in the HF and low fat diet fed mice. Serum levels of glucose and triglycerides were significantly lower in mice given 0.5% SE and 1% SE than those in the HF mice but was not significantly lower than those in the LF diet fed mice. The mice fed the HF diet had a significant elevation in blood glucose level throughout the oral glucose tolerance test compared with those in the LF, 0.5% SE, and 1% SE groups (**Figure 16**). The AUC for glucose decreased by 23% and 25% in mice given 0.5% SE or 1% SE, respectively, compared to that of the HF mice (**Figure 16** and **Table 19**). Serum insulin levels were significantly lower in mice fed 1% SE or the LF diet than in mice in the HF and 0.5% SE groups. No significant differences in levels of serum HDL-cholesterol, GOT, and GPT were observed among the groups.

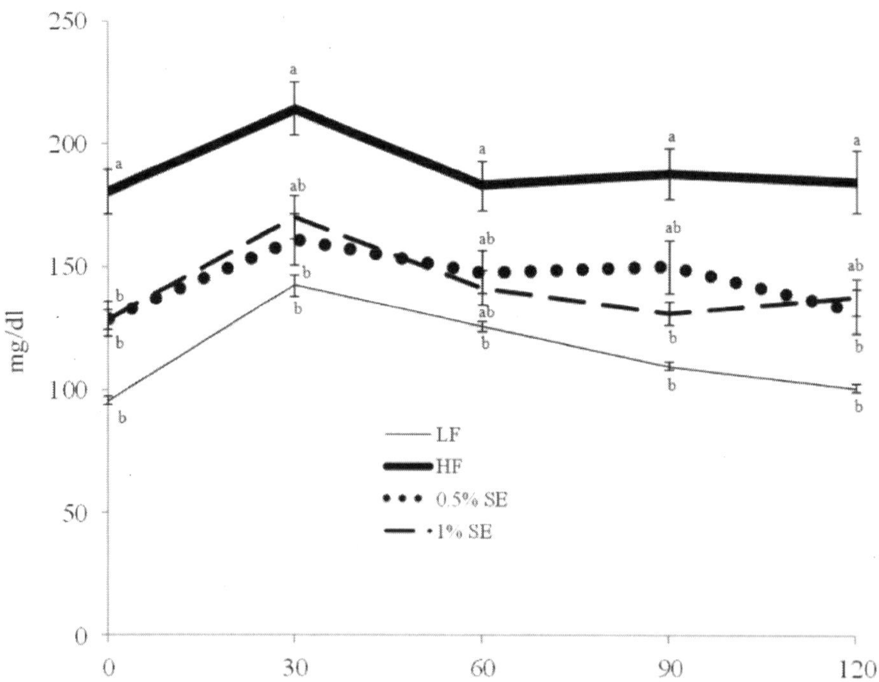

Figure 16: Effect of sorghum extract on the oral glucose tolerance test. Values are expressed as mean ± standard error; n = 7/group. LF, low fat diet; HF, high fat diet; 0.5% SE, high fat diet with 0.5% of dietary intake as sorghum extract, 1% SE, high fat.

Table 19: Lipid profiles, liver function, and levels of glucose and insulin in serum

	LF	HF	0.5% SE	1% SE
Triglycerides (mmol/L)	0.90 ± 0.11^{a2}	1.27±0.10^{b}	0.80 ± 0.13^{b}	0.96 ± 0.13^{b}
Total cholesterol (mmol/L)	4.09 ± 0.17	4.31 ± 0.20^{b}	3.63 ± 0.36^{ac}	3.37 ± 0.22^{c}
HDL-cholesterol (mmol/L)	2.08 ± 0.10	2.27 ± 0.09	2.09 ± 0.21	2.00 ± 0.08
LDL-cholesterol (mmol/L)	1.62 ± 0.06^{a}	1.79 ± 0.09^{a}	1.17 ± 0.16^{b}	0.88 ± 0.09^{b}
GOT (IU/L)	34.62 ± 2.90	44.71 ± 3.78	40.15 ± 6.27	39.92 ± 5.56
GPT (IU/L)	12.53 ± 1.18	24.62 ± 5.74	17.26 ± 3.74	16.26 ± 2.36
Glucose (mmol/L)	5.29 ± 3.39^{a}	10.01± 1.03^{b}	7.14 ± 0.77^{a}	7.14 ± 0.43^{a}
Insulin (pmol/L)	59.53 ± 12.21^{a}	120.58 ± 13.75^{b}	103.79 ± 27.47^{b}	59.53 ± 12.21^{a}
AUC of glucose (mmol min/L)	790.63 ± 23.98^{a}	1287.07 ± 137.95^{b}	979.75± 123.75^{a}	957.19 ± 80.65^{a}

Effect of sorghum extract on antidiabetic protein expression

PPAR-γ expression was significantly higher in mice given 1% SE than in the HF and 0.5% SE groups but was similar to that in the LF diet fed mice (**Figure 17**). TNF-α expression was significantly lower in mice given 1% SE than in the HF and LF diet fed mice and was significantly lower in mice given 0.5% SE than in the LF diet fed mice. Adiponectin expression was significantly higher in mice given 0.5% and 1% SE than that in the HF diet fed mice.

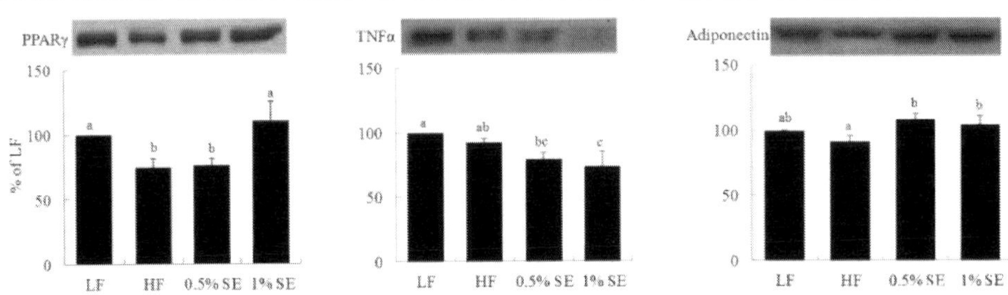

Figure 17: Effect of sorghum extracts on peroxisome proliferator-activated receptor gamma (PPAR-γ), tumor necrosis factor-α (TNF-α), and adiponectin protein expression in adipose tissue. Values are expressed as means ± standard error.

As hypothesized, the SE appeared to have induced antidiabetic effects in mice fed a HF diet through a mechanism that increased adiponectin and decreased TNF-α via overexpression of PPAR-γ, leading to improved insulin sensitivity. Administration of both 0.5% and 1% SE significantly reduced serum glucose levels, but only the 1% SE treatment significantly lowered serum insulin level, suggesting that the 1% concentration increased antidiabetic effects, although our findings were not dose-dependent.

These findings are consistent with a recent study that showed that SE at doses of 250 and 500 mg/kg of body weight significantly decrease serum glucose levels in streptozotocin-induced diabetic rats (Chung et al., 2011). Kim et al. (2011) showed that SE is rich in polyphenols, which strongly inhibit *in vitro* α-glucosidase and α-amylase activities, and are targets for the development of diabetic drugs. Studies have also reported that polyphenols from cocoa extract (Ruzaidi et al., 2005) and *Salacia reticulate* extract (Akase et al., 2011) reduce serum glucose levels in type 1 diabetic rats. However, Lakshmi et al. (Lakshmi and Vimala, 1996) observed that consumption of whole sorghum significantly lowers fasting level and the glucose AUC in type 2 diabetic subjects, suggesting that these results could be due to fiber. Dietary fiber lowers blood glucose level by delaying gastric emptying, intestinal transit time, and carbohydrate absorption (Ray et al., 1983); however, we previously reported that consuming sorghum grain has no effect on serum glucose or insulin levels compared with those in hyperlipidemic rats fed white rice (Lee et al., 2010). These inconsistent results may be due to the type of sorghum consumed, sorghum grain vs. extract, or the animal model used.

PPAR-γ is involved in glucose uptake and lipid storage in adipocytes (Chawla et al., 1994; Tontonoz et al., 1994). Activation of PPAR-γ to improve the sensitivity of insulin receptors is the predominant mechanism for the antidiabetic efficacy of PPAR-γ agonists (Willson et al., 2000). Thiazolidinedione, a synthetic PPAR-γ ligand, significantly increases insulin sensitivity in patients with diabetes, interferes with TNF-α activity, and increases adiponectin levels (Arner, 2003). Adipocytokines, primarily adiponectin and TNF-α released from adipocytes, appear to be associated with glucose metabolism (Thamer et al., 2002). Therefore, the hypoglycemic effect of sorghum in this study could be explained by increased PPAR-γ expression in adipose tissue, which consequently increased adiponectin expression and decreased TNF-α expression.

A significant decrease in total cholesterol, LDL-cholesterol, and triglyceride levels was observed in the SE-treated groups. Carr et al. (Carr et al., 2005) also reported that a sorghum lipid extract from whole kernels exerts a cholesterol-lowering effect by reducing cholesterol absorption with a concomitant increase in fecal sterol excretion. Polyphenols from green tea reduce plasma LDL-cholesterol concentrations by inhibiting cholesterol synthesis (Bursill et al., 2007; Singh et al., 2005; Maron, et al., 2003). Because the cholesterol-lowering effects can be explained by decreased cholesterol synthesis and/or increased cholesterol excretion, the effect of SE on plasma cholesterol concentration in this study may have been due to either increased cholesterol excretion in the feces or decreased cholesterol synthesis in the liver. Chung et al. (Chung et al., 1972) also reported that a SE reduces plasma triglyceride concentrations in rats with diet-induced obesity, but the hyperlipidemic mechanism is unknown.

A few limitations of this study should be mentioned. No active control group was used (e.g., mice taking antidiabetic medication), dose dependency was not clearly detected due to the doses selected, and the hypocholesterolemic metabolism was not determined. We are currently conducting an animal study to examine the cholesterol-lowering action mechanism of sorghum. In conclusion, our results demonstrate that a SE appeared to have an antidiabetic effect and may have exerted its therapeutic effects through PPAR-γ overexpression in mice fed a HF diet. This is the first study to characterize the antidiabetic mechanisms of sorghum. However, future studies should examine the hypoglycemic effect of sorghum on protein expression during hepatic glucose production and glucose uptake by skeletal muscle.

Cereal crops have recently experienced increased interest due to their potential health benefits. It has been suggested that the intake of whole grain foods is beneficial to the prevention and management of diabetes mellitus. In this study, we investigated the inhibitory effect of 70% EtOH extracts from different cultivars of sorghum, foxtail millet and proso millet on α-glucosidase and α-amylase. Among the six sorghum cultivars, Mongdang-susu(SS-1), Me-susu(SS-2), Susongsaengi-susu(SS-3) and Sikyung-susu(SS-4) extracts exhibited higher inhibitory activities against α-glucosidase (IC50 = 1.1–1.4 lg/ml) than acarbose, reference inhibitor (IC50 = 2.1 lg/ml). In addition, these extracts strongly inhibited degradation of starch by pancreatic and salivary α-amylase, whereas extracts from foxtail and proso millets exhibited no visible or detectible inhibitory effect on α-amylase or on α-glucosidase activity. These *in vitro* studies indicate the potential of sorghum in the development of effective anti-diabetic agents. In conclusion, our investigation indicated that sorghum, the fifth most important cereal in the world, is an effective inhibitor of glucosidase, with inhibitory activity against B. stearothermophilus α-glucosidase as well as porcine pancreatic and human salivary α-amylases. Although the inhibitory effect of sorghum extracts has been established in vitro, these findings indicate that sorghum extract may potentially suppress the increase in post-prandial glucose by starch (Ju-Sung Kim et al., 2011).

Glycemic responses of diabetic subjects fed on various recipes

Consumption of whole sorghum recipes showed significantly lower (P<O.O5) peak levels of plasma glucose, lesser per cent peak rise, AUC and lesser per cent glycemic response in diabetics than consumption of dehulled sorghum recipes and other recipes prepared with wheat and rice. These differences may be due to the difference in fiber content of the recipes and also due to the differences in the rate of digestion, absorption and utilisation of carbohydrates (Kamini Devi, 1972). Number of studies has been documented on the positive effects of fiber in reducing the plasma glucose levels (Bose, 1979; Jenkins et al., 1980; Jenkins 1981; Hollnbeck, et al., 1985; Kiehm et al., 1976). The role of fiber in reducing the plasma glucose levels has been attributed to slower gastric emptying or formation of unabsorbable complexes with available carbohydrates in gut lumen.

Whole sorghum recipes namely missiroti, (calculated upma and dhokla value - 4.8 g, 4.lg 2.2g) had comparatively higher fiber content in comparision with the dehulled sorghum (2.7g, 2.5g, 1.8g) and other recipes (3.0g, l.0g, 0.7g). The beneficial effect of fiber in diabetics could be attributed to its effect on increased efficiency of endogenous insulin to metabolize glucose. Though there is not much difference in the fiber content of whole sorghum missiroti and wheat missiroti, the consumption of whole sorghum missiroti resulted in lower plasma glucose levels. There are also evidences to show that wheat bran is effective only when it is given in large quantities (Harold et al., 1985) and also given for longer periods of time (Osluzzi and Crivellent, 1978; Vaaler and Honssen, 1983; Villaumet et al., 1984). However, the result of the present study showed that wheat bran is also effective in reducing the plasma glucose levels, but not as effective as whole sorghum. Among the three whole sorghum recipes tested, consumption of whole sorghum semolina upma resulted in lower glycemic responses, as compared to the other whole sorghum recipes. The flour form of sorghum used in the preparation of missiroti and dhokla makes the carbohydrate definitely more accessible to the enzymes against the grit form of the sorghum used in upma. These results are in conformation with the findings of 0' Dea et al. (1980). Dehulled sorghum recipes are also not as effective as whole sorghum recipes in reducing the plasma glucose levels which may be due to the loss of bran and also the protein during dehulling (Shanaaz Saldana 1991). The results of the present study suggest that the use of whole sorghum recipes has a smoothening effect on post-prandial plasma glucose levels. It may not be practicable to completely replace the cereal in a diabetic diet with sorghum. Whole sorghum can be substituted in a meal where the plasma glucose levels tend to be the maximum during the day. A combination of cereal and millet may also be beneficial to counteract the hypoglycemic effect of a cereal such as rice or wheat. Indiscriminate long term use of millet fibre might affect the bioavailability of minerals both in normal and diabetic patients. Hence, a critical level of millet fibre which does not affect the bioavailability of minerals should be arrived at before making any recommendation.

5. Celiac Disease

Although it is a relatively "new" condition, it seems to be gaining awareness all around the world. Celiac disease is a severe allergy to gluten, which is primarily found in wheat-based products. Surprisingly, wheat/gluten is found in thousands of normal food items, making life for those suffering from Celiac disease very difficult. Fortunately, alternative grains and grasses, such as sorghum, can be eaten safely by those suffering from this increasingly common condition, without the painful inflammation, nausea, and gastrointestinal damage that gluten causes.

Background and aims of celiac disease

Celiac disease is a condition in which genetically predisposed people have an autoimmune reaction to gluten proteins found in all wheat types and closely related cereals such as barley and rye. This reaction causes the formation of autoantibodies and the destruction of the villi in the small intestine, which results in malabsorption of nutrients and other gluten-induced autoimmune diseases. Sorghum is a cereal grain with potential to be developed into an important crop for human food products. The flour produced from white sorghum hybrids is light in colour and has a bland, neutral taste that does not impart unusual colours or flavours to food products. These attributes make it desirable for use in wheat-free food products. While sorghum is considered as a safe food for celiac patients, primarily due to its relationship to maize, no direct testing has been conducted on its safety for gluten intolerance. Therefore studies are needed to assess its safety and tolerability in celiac patients. Thus the aim of the present study was to assess safety and tolerability of sorghum flour products in adult celiac disease patients, utilizing an *in vitro* and *in vivo* challenge.

Sorghum protein digests did not elicit any morphometric or immunomediated alteration of duodenal explants from celiac patients. Patients fed daily for 5 days with sorghum-derived food product did not experience gastrointestinal or non-gastrointestinal symptoms and the level of anti-transglutaminase antibodies was unmodified at the end of the 5-days challenge. Sorghum-derived products did not show toxicity for celiac patients in both *in vitro* and *in vivo* challenge. Therefore sorghum can be considered safe for people with celiac disease (Carolina Ciacci et al., 2007).

6. Bone Health

Magnesium is found in high quantities in sorghum, which means that your calcium levels will be properly maintained, as magnesium increases calcium absorption in the body. These two minerals are also integral to the development of bone tissue and speeds up the healing of damaged or aging bones. This can prevent conditions like osteoporosis and arthritis, thereby keeping you active and healthy well into your old age.

The chemical composition of biofuels has not received adequate attention given that it is an important aspect in the introduction of energy crops. In this study, the ash content and mineral composition (C, N, Al, Ca, Cl, Fe, K, Mg, Na, P, S, and Si) of stems, leaves and reproductive organs of some promising energy crops were determined and compared with the respective recommended thresholds reported in literature. Overall, cynara exhibited the highest ash and mineral contents, which indicate high slagging, fouling and corrosion tendencies. However, cynara also showed the lowest Si content, both in leaves (4.3 g kg^{-1}) and in stems (0.9 g kg^{-1}). Sweet sorghum and giant reed exhibited the highest N content (up to 16 g kg^{-1}), which greatly exceeded the recommended limits in leaves. Importantly, Cl always exceeded the recommended limits (up to 18 mg kg^{-1} in cynara), both in stems and in leaves, thus resulting in a major stumbling block for all crops. Several significant correlations among elements were found at a single plant part; conversely these correlations were generally very weak considering different plant components, with the exception of K ($r = 0.91^{**}$),

P (r = 0.94**) and ashes (r = 0.64**). Generally, leaves resulted in a significant deterioration of biofuel quality when compared with stems and flower heads. Therefore, agricultural strategies aimed at reducing the leaf component (e.g. by delaying the harvest) may considerably improve the suitability of biofuels for current combustion plants.

Cynara showed the highest ash and mineral contents, thus resulting in significant slagging, fouling and corrosion tendencies. Giant reed and sweet sorghum are known to be capable of providing high biomass yields, yet they also showed considerable N and Si contents, especially in leaves and reproductive organs. In general, all crops showed high Cl and S contents, which are strongly associated with corrosion and HCl emissions. Again, the highest Cl content was found in cynara, which suggests that special measures have to be taken for processing this biofuel. Switchgrass and miscanthus showed the overall better biomass quality. In all crops, stems were much better than leaves and reproductive organs, and in most cases they showed lower than recommended mineral concentrations. Therefore, agricultural strategies aimed at pursuing primary sources with a preponderant stem component should be recommended (Andrea Monti et al., 2008).

7. Circulation and Red Blood Cell Development

Copper and iron are also found in sorghum, and in a similar way as magnesium and calcium, copper helps to increase the uptake and absorption of iron into the body. This means a decreased likelihood of developing anemia, which is another name for iron deficiency. With enough iron and copper in your system, red blood cell development is increased, thereby boosting circulation of the blood, stimulating cellular growth and repair, and increasing the hair growth of the scalp, while also boosting energy levels throughout the body. A single serving of sorghum contains 58% of your daily recommended intake of copper.

This research study describes an experiment, carried out under controlled environment conditions, to investigate the effects of a fulvic acid fraction of soil organic matter on growth, cadmium (Cd) uptake and redistribution by sorghum. In addition the uptake of copper (Cu), zinc (Zn) and iron (Fe) was also determined. Sorghum was grown in nutrient solutions with 0, 0.1, 1 and 10 mg Cd dm^{-3}, in the absence and presence of organic matter (32 mg C dm^{-3}), for various periods up to 20 days. A decrease in sorghum biomass due to Cd toxicity was observed at 10 mg Cd dm^{-3}, but for concentrations of 0.1 and 1 mg Cd dm^{-3} the biomass was increased compared with control, without visual toxicity symptoms. The presence of organic matter (OM) further increased biomass production. Cadmium was mainly retained in sorghum roots, as usually found in tolerant plants, but Cd accumulation in sorghum was greater than in other *Gramineae*, or even more tolerant plants such as lettuce. The presence of OM decreased the bioavailability of Cd that was partially retained in solution by the OM ligands. However, OM promoted the translocation of Cd to shoots, an effect that may pose a risk to public health because plant-animal transfer of Cd could be enhanced. The presence of OM decreased the uptake of Cu, Zn and Fe. The presence (vs. absence) of 0.1 mg Cd dm^{-3} enhanced thc uptake of Fe, both in the absence and presence of OM (Pinto et al., 2004).

8. Energy Levels

Niacin, also known as vitamin B_6 is a key component in transforming food into usable energy and fuel for the body. Breaking down and metabolizing nutrients into energy will keep your energy levels stable throughout the day, rather than the cheap peaks and troughs that sugar boosts give you? Sorghum contains 28% of your recommended niacin intake per day.

Two varieties of *Sorghum vulgare*, milo and jowar, have been studied. Samples cooked at neutral pH gave total niacin values of 46 and 29 mg/kg, respectively, on an air-dry basis. In extracts made from these at neutral pH and fractionated after gel-filtration, most of the niacin was in high molecular weight (i.e., bound) forms. In a bioassay for niacin, the boiled milo gave a growth response equivalent to 34% of its total niacin being available to young rats. Sorghum has behaved similarly to other mature cereal grains in these tests of niacin availability (Carter and Carpenter, 1981).

9. A Final Word of Caution

There are no known concerns or cautions against adding sorghum to your diet. As it is belong to the grass categories, there is the possibility of some people being allergic to it, but the cases of sorghum being allergenic are very rare. Furthermore, with the high content of certain minerals and vitamins, the only real danger is getting too much of a good thing, so eat sorghum in moderation and enjoy all of the wonderful health benefits.

REFERENCES

Abdel-Aal, E.M., Hucl, P. (2003). Composition and stability of anthocyanins in blue-grained wheat. Journal of Agricultural and Food Chemistry 51, 2174–2180.

Adamson, G.E., Lazarus, S.A., Mitchell, A.E., Prior, R.L., Cao, G., Jacobs, P.H., Kremers, B.G., Hammerstone, J.F., Rucher, R.B., Ritter, K.A., Schmidtz, H.H. (1999). HPLC method for the quantification of procyanidins in cocoa and chocolate samples and correlation to total antioxidant capacity. Journal of Agricultural and Food Chemistry 47, 4184–4188.

Adom, K.K., Liu, R.H. (2002). Antioxidant activity of grains. Journal of Agricultural and Food Chemistry 50, 6182–6187.

Agullo, E., Rodriguez, M.S. (1995). Evaluation of 3 rapid methods to determine tannins in sorghum grains. Anales de la Asociaci_on Qu_imica Argentina 83, 83–87.

Akase T, Shimada T, Harasawa Y, Akase T, Ikeya Y, Nagai E, Iizuka S, Nakagami G, Iizaka S, Sanada H, and Aburada M. (2011). Preventive effects of Salacia reticulata on obesity and metabolic disorders in TSOD mice. Evid Based Complement. Alternat Med. 484-590.

Akingbala, J.O. (1991). Effect of processing on flavonoids in millet (Pennisetum americanum) flour. Cereal Chemistry 68, 180–183.

Almadal, T. P., and Vilstrup, H. (1988). Strict insulin treatment normalizes the organic nitrogen contents and the capacity of urea-N synthesis in experimental diabetes in rats. Diabetologica, 31, 114–118.

Al-Mamary, M., Al-Habori, M., Al-Aghbari, A., Al-Obeidi, A. (2001). In vivo effects of dietary sorghum tannins on rabbit digestive enzymes and mineral absorption. Nutrition Research 21, 1393–1401.

Ambula, M.K., Oduho, G.W., Tuitoek, J.K. (2003). Effects of high-tannin sorghum and bentonite on the performance of laying hens. Tropical Animal Health and Production 35, 285–292.

Ambula, M.K., Oguho, G.W., Tuitoek, J.K. (2001). Effects of sorghum tannins, a tannin binder (polyvinylpyrrolidone) and sorghum inclusion level on the performance of broiler chicks. Asian-Australian Journal of Animal Science 14, 1276–1281.

Anderson, J.W. (2003). Whole grain protects against atherosclerotic cardiovascular disease. Proceedings of the Nutrition Society 62, 135–142.

Andrea Monti, Nicola Di Virgilio, and Gianpietro Venturi. (2008). Mineral composition and ash content of six major energy crops. Biomass and Bioenergy, 32: 216 – 223.

Andreasen, M.F., Christensen, L.P., Meyer, A.S., Hansen, A.S. (2000). Contents of phenolic acids and ferulic acid dehydrodimers in 17 rye (*Secale cereale* L.) varieties. Journal of Agricultural and Food Chemistry 48, 2837–2842.

Andreasen, M.F., Kroon, P.A., Williamson, G., Garcia-Conesa, M. (2001a). Esterase activity able to hydrolyze dietary antioxidant hydroxycinnamates is distributed along the intestine of mammals. Journal of Agricultural and Food Chemistry 49, 5679–5684.

Andreasen, M.F., Kroon, P.A., Williamson, G., Garcia-Conesa, M. (2001b). Intestinal release and uptake of phenolic antioxidant diferulic acid. Free Radical Biology and Medicine 31, 304–314.

Antony, U., Chandra, T.S. (1998). Antinutrient reduction and enhancement in protein, starch, and mineral availability in ferment flour of finger millet (Eleusine coracana). Journal of Agricultural and Food Chemistry 46, 2578–2582.

Anwar, M. M., and Meki, A. M. R. (2003). Oxidative stress in streptozotocininduced diabetic rats: Effects of garlic oil and melatonin. Comparative Biochemistry and Physiology, 135, 539–547.

Armstrong, W.D., Featherston, W.R., Rogler, J.C. (1974). Effects of bird resistant sorghum grain and various commercial tannins on chick performance. Poultry Science 53, 2137–2142.

Arner P. (2003). The adipocyte in insulin resistance: key molecules and the impact of the thiazolidinediones.Trends Endocrinol Metab. 14:137–145.

Astley, S.B. (2003). Dietary antioxidants – past, present and future? Trends in Food Science and Technology 14, 63–98.

Audilakshmi, S., Stenhouse, J.W., Reddy, T.P., Prasad, M.V.R. (1999). Grain mould resistance and associated characters of sorghum genotypes. Euphytica 107, 91–103.

Avato, P., Bianchi, G., Murelli, C. (1990). Aliphatic and cyclic lipid components of Sorghum plant organs. Phytochemistry 29, 1073–1078.

Awika, J. M. (2000). Sorghum phenols as antioxidants. M.S. Thesis, Texas AandM University: College Station, TX.

Awika, J.M. (2003). Antioxidant properties of sorghum. Ph.D. Dissertation, Texas A and M University, College Station, TX. 248 L. Dykes, L.W. Rooney / Journal of Cereal Science 44 (2006) 236–251

Awika, J.M., Dykes, L., Gu, L., Rooney, L.W., Prior, R.L. (2003a). Processing of sorghum (Sorghum bicolour) and sorghum products alters procyanidin oligomer and polymer distribution and content. Journal of Agricultural and Food Chemistry 51, 5516–5521.

Awika, J.M., McDonough, C.M., Rooney, L.W. (2005). Decorticating sorghum to concentrate healthy phytochemicals. Journal of Agricultural and Food Chemistry 53, 6230–6234.

Awika, J.M., Rooney, L.W. (2004). Sorghum phytochemicals and their potential impact on human health. Phytochemistry, 1199–1221.

Awika, J.M., Rooney, L.W., Waniska, R.D. (2004a). Anthoycanins from black sorghum and their antioxidant properties. Food Chemistry 90, 293–301.

Awika, J.M., Rooney, L.W., Waniska, R.D. (2004b). Properties of 3-deoxyanthocyanins from sorghum. Journal of Agricultural and Food Chemistry 52, 4388–4394.

Awika, J.M., Rooney, L.W., Wu, X., Prior, R.L., Cisneros-Zevallos, L. (2003b). Screening methods to measure antioxidant activity of sorghum (Sorghum bicolour) and sorghum products. Journal of Agricultural and Food Chemistry 51, 6657–6662.

Bacon, J.R., Rhodes, M.J.C. (1998). Development of a competitive assay for the evaluation of of the binding of human parotid salivary proteins to dietary complex phenols and tannins using a eroxidase-labelled tannin. Journal of Agricultural and Food Chemistry 46, 5083–5088.

Barham, D., and Trinder, P. (1972). An improved colour reagent for the determination of blood glucose by the oxidase system. The Analyst, 97(1151), 142–145.

Bate-Smith, E.C. (1969). Luteoforol (30, 4, 40, 5, 7-pentahydroxyflavan) in Sorghum vulgare L. Phytochemistry 8, 1803–1810.

Bate-Smith, E.C., Rasper, V. (1969). Tannins of grain sorghum: Luteoforol (leucoluteolinidin), 30,4,40,5,7-pentahydroxyflavan. Journal of Food Science 34, 203–209.

Beta, T., Rooney, L.W., Marovatsanga, L.T., Taylor, J.R.N. (1999). Phenolic compounds and kernel characteristics of Zimbabwean sorghums. Journal of the Science of Food and Agriculture 79, 1003–1010.

Beta, T., Rooney, L.W., Marovatsanga, L.T., Taylor, J.R.N. (2000a). Effect of chemical treatments on polyphenols and malt quality in sorghum. Journal of Cereal Science 31, 295–302.

Beta, T., Rooney, L.W., Taylor, J.R.N. (2000b). Effect of chemical conditioning on the milling of high-tannin sorghum. Journal of the Science of Food and Agriculture 80, 2216–2222.

Blakeley, M.E., Rooney, L.W., Sullins, R.D., Miller, F.R. (1979). Microscopy of the pericarp and the testa of different genotypes of sorghum. Crop Science 19, 837–842.

Blessin, C.W., Vanetten, C.H., Dimler, R.J. (1963). An examination of anthocyanogens in grain sorghums. Cereal Chemistry 40, 241–250.

Boren, B., Waniska, R.D. (1992). Sorghum seed colour as an indicator of tannin content. Journal of Applied Poultry Research 1, 117–121.

Bors, W., Michel, C., Stettmaier, K. (2000). Electron paramagnetic resonance studies of radical species of proanthocyanidins and gallate esters. Archives of Biochemistry and Biophysics 374, 347–355.

Bose T, (1979). High dietary fibre in routine Bengali diets - a check diabetes mellitus. Ind. J.Nutr. Diet. 1979: 16:312-315.

Boveris, A.D., Galatro, A., Sambrotta, L., Ricco, R., Gurni, A.A., Puntarulo, S. (2001). Antioxidant capacity of 3-deoxyanthocyanidin from soybean. Phytochemistry 58, 1097–1105.

Boyd, W. (2000). Natural colours as functional ingredients in healthy foods. Cereal Foods World, 45(5), 221–222.

Brandon, M.J., Foo, L.Y., Porter, L., Meredith, P. (1982). Proanthocyanidins of barley and sorghum; composition as a function of maturity of barley ears. Phytochemistry 12, 2953–2957.

Bridle, P., and Timberlake, C. F. (1997). Anthocyanins as natural food colours – selected aspects. Food Chemistry, 58, 103–109.

Bridle, P., Timberlake, C.F., 1997. Anthocyanins as natural food colours-selected aspects. Food Chemistry 58, 103–109.

Bronnun-Hansen, K., Jacobsen, F., and Flink, J. M. (1985). Anthocyanin colourants from elderberry (*Sambucus nigra* L.) 1. Process consideration of production of liquid extract. Journal of Food Technology, 20, 703–711.

Bronnun-Hansen, K., Jacobsen, F., Flink, J.M. (1985). Anthocyanin colourants from elderberry (*Sambucus nigra* L.). 1. Process consideration of production of liquid extract. Journal of Food Technology 20, 703–711.

Brouillard, R., Wigand, M. C., Dangles, O., and Cheminat, A. (1991). pH and solvent effects on the co-pigmentation reaction of malvin with polyphenols, purine and pyrimidine-derivatives. Journal of the Chemical Society, Perkin Transactions, 2(8), 1235–1241.

Bullard, R.W., Gebrekidan, B. (1989). Agronomic techniques to reduce quelea damage to cereals. In: Bruggers, R.L., Elliott, C.C.H. (Eds.), Quelea quelea Africa's bird pest. Oxford University Press, New York, NY, pp. 281–292.

Bullard, R.W., York, J.O. (1985). Breeding for bird resistance in sorghum and maize. In: Russell, G.E. (Ed.), Plant Breeding Progress Reviews, vol. 1. Butterworths, Surrey, pp. 193–222.

Bullard, R.W., York, J.O. (1996). Screening grain sorghums for bird tolerance and nutritional quality. Crop Protection 15, 159–165.

Bunger, W.B., Kummerow, F.A. (1951). A comparison of several methods for the separation of unsaponifiable material from carnauba and grain sorghum waxes. Journal of the American Oil Chemists Society 28, 121–123.

Burns, R.E. (1971). Methods for estimation of tannin in grain sorghum. Agronomy Journal 63, 511.

Bursill CA, Abbey M, and Roach P. D. (2007). A green tea extract lowers plasma cholesterol by inhibiting cholesterol synthesis and upregulating the LDL receptor in the cholesterol-fed rabbit. Atherosclerosis. 193:86–93.

Butler, L.G. (1982). Relative degree of polymerization of sorghum tannin during seed development and maturation. Journal of Agricultural and Food Chemistry 30, 1090–1094.

Butler, L.G. (1989). Sorghum polyphenols. In: Cheeke, P.R. (Ed.), Toxicants of Plant Origin, Phenolics, vol. IV. CRC Press, Inc., Boca Raton, FL, pp. 95–121.

Butler, L.G., Rogler, J.C. (1992). Biochemical mechanisms of the antinutritional effects of tannins. In: Ho, C.-T., Lee, C.Y., Huang, M.-T. (Eds.), Phenolic Compounds in Food and Their Effects on Health. I., ACS Symposium Series, vol. 506. American Chemical Society, Washington, DC, pp. 298–304.

Bvochora, J.M., Danner, H., Miyafuji, H., Braun, R., Zvauya, R. (2005). Variation of sorghum phenolic compounds during the preparation of opaque beer. Process Biochemistry 40, 1207–1213.

Carlson, J. J., Eisenmann, J. C., Norman, G. J., and Ortiz, K. A. (2011). Young. Dietary fiber and nutrient density are inversely associated with the Metabolic Syndrome in US ado lescents. JADA, 111; 1688.

Carmona, A., Borgudd, L., Borges, G., Levy-Benshimol, A. (1996). Effects of black bean tannins on in vitro carbohydrate digestion and absorption. Journal of Nutritional Biochemistry 7, 445–450.

Carolina Ciacci, Luigi Maiuri, Nicola Caporaso, Cristina Bucci, Luigi Del Giudice, Domenica Rita Massardo, Paola Pontieri, Natale Di Fonzo, Scott R. Bean, Brian Ioerger and Marco Londei, (2007). Celiac disease: *In vitro* and *in vivo* safety and palatability of wheat-free sorghum food products. Clinical Nutrition, 26 (6): 799-805.

Carr TP, Weller CL, Schlegel VL, Cuppett SL, Guderian DM, Jr and Johnson KR. (2005). Grain sorghum lipid extract reduces cholesterol absorption and plasma non-HDL cholesterol concentration in hamsters. J Nutr. 135:2236–2240.

Carter, E. G. A. and Carpenter, K. J. (1981). Bound niacin in sorghum and its availability. Nutrition Research, 1(6): 571-579.

Castano, G., Menendez, R., Mas, R., Amor, A., Fernandez, J.L., Gonzalez, R.L., Alvarez, E. (2002). Effects of lovastatin on lipid profile and lipid peroxidation in patients with dyslipidemia associated with type 2 diabetes mellitus. International Journal of Clinical Pharmacolology Research 22 (3–4), 89–99.

Cedillo-Sebastian, G. (2005). Nutraceutical tortillas and tortilla chips prepared with bran from specialty sorghums. M.S. Thesis, Texas AandM University, College Station, TX.

Cevallos-Casals, B.A., Cisneros-Zevallos, L. (2003). Stoichiometric and kinetic studies of phenolic antioxidants from Andean purple corn and red-fleshed sweet potato. Journal of Agricultural and Food Chemistry 51, 3313–3319.

Chang, S.I., Fuller, H.L. (1964). Effect of tannin content of grain sorghums on their feeding value for growing chicks. Poultry Science 43, 30–36.

Chawla A, Schwarz EJ, Dimaculangan DD, and Lazar MA. (1994). Peroxisome proliferator-activated receptor (PPAR) gamma: adiposepredominant expression and induction early in adipocyte differentiation.Endocrinology. 135:798–800.

Chen, F., Cole, P., Mi, Z., Xing, L.-Y. (1993). Corn and wheat-flour consumption and mortality from esophageal cancer in Shanxi, China. International Journal of Cancer 53, 902–906.

Chibber, B.A.K., Mertz, E.T., Axtell, J.D. (1978). Effects of dehulling on tannin content, protein distribution, and quality of high and low tannin sorghum. Journal of Agricultural and Food Chemistry 26, 679–683.

Cho, S.-H., Choi, Y., Ha, T.-Y. (2000). In vitro and in vivo effects of prosomillet, buckwheat and sorghum on cholesterol metabolism. FASEB Journal 14, A249.

Chung IM, Kim EH, Yeo MA, Kim SJ, Seo MC, and Moon HI. (2011). Antidiabetic effects of three Korean sorghum phenolic extracts in normal and streptozotocin-induced diabetic rats. Food Res Int. 44:127–132.

Chung IM, Yeo MA, Kim SJ, Kim MJ, Park DS, and Moon HI. (2011). Antilipidemic activity of organic solvent extract from Sorghum bicolour on rats with diet-induced obesity. Hum Exp Toxicol. 30:1865–1868.

Chung, K.-T., Tit, Y.W., Cheng, I.W., Yao-Wen, H., Yuan, L. (1998). Tannins and human health: a review. Critical Reviews in Food Science and Nutrition 38 (6), 421–464.

Clifford, M. N. (2000). Anthocyanins – nature, occurrence and dietary burden. Journal of the Science of Food and Agriculture, 80, 1063–1072.

Cousins, B.W., Tanksley, T.D., Knabe, D.A., Zebrowska, T. (1981). Nutrient digestibility and performance of pigs fed sorghums varying in tannin concentration. Journal of Animal Science 53, 1524–1529.

Cummings, D.P, Axtel, J.D. (1973). Relationships of pigmented testa to nutritional quality of sorghum grain in inheritance and improvement of protein quality in sorghum. Research Progress Report, Purdue University, West Lafayette, IN, p. 112.

Dalton, J.L., Mitchell, H.L. (1959). Fractionation of grain sorghum wax. Journal of Agricultural and Food Chemistry 7, 570–573.

Deprez, S., Brezillon, C., Rabot, S., Philippe, C., Mila, I., Lapierre, C., Scalbert, A. (2000). Polymeric proanthocyanidins are catabolized by human colonic microflora into low-molecular weight phenolic acids. Journal of Nutrition 130, 2733–2738.

Deprez, S., Mila, I., Huneau, J.F., Tome, D., Scalbert, A. (2001). Transport of proanthocyanidin dimer, trimer, and polymer across monolayers of human intestinal epithelial Caco-2 cells. Antioxidants and Redox Signaling 3, 957–967.

Deshpande, S.S., Cheryan, M. (1986). Tannin analysis of food products. CRC Critical Reviews in Food Science and Nutrition 24, 401–449.

Deshpande, S.S., Cheryan, M., Salunkhe, D.K. (1985). Evaluation of vanillin assay for tannin analysis of dry beans. Journal of Food Science 50, 905–910.

Dewar, J., Orovan, E., Taylor, J.R.N. (1997). Effect of alkaline steeping on water uptake and malt quality in sorghum. Journal of the Institute of Brewing 103, 283–285.

Dicko, M.H., Gruppen, H., Traore, A.S., Van Berkel, W.J.H., Voragen, A.G.J. (2005). Evaluation of the effect of germination on phenolic compounds and antioxidant activities in sorghum varieties. Journal of Agricultural and Food Chemistry 53, 2581–2588.

Dube, A., Bharti, S., and Laloraya, M. M. (1992). Inhibition of anthocyanin synthesis by cobaltous ions in the 1st internode of Sorghum-bicolour L. Moench. Journal of Experimental Botany, 43(255), 1379–1382.

Duke, J.A. (1992). Biological Active Phytochemicals and Their Activities. CRC Press, Inc., Boca Raton, FL.

Dunford, N.T., King, J.W. (2000). Phytosterol enrichment of rice bran by a supercritical carbon dioxide fraction technique. Journal of Food Science 65, 1395–1399.

Duodu, K. G., Nunes, A., Delgadillo, I., Parker, M. L., Mills, E. N. C., and Belton, P. S. (2002). Effect of grain structure and cooking on sorghum and maize in vitro protein digestibility. Journal of Cereal Science, 35, 161–174.

Dykes, L., and Rooney, L. W. (2006). Journal of Cereal Science 44: 236–251.

Dykes, L., Awika, J.M., McDonough, C.M., Rooney, L.W., Waniska, R.D. (2002). False positives for tannin sorghum in non-tannin sorghum using the bleach test. AACC Annual Meeting Abstracts. URL:/http://www.aaccnet.org/meetings/2002/ abstracts/a02ma286. aspS.

Dykes, L., Rooney, L.W., Waniska, R.D. (2004). Normal-phase HPLC analysis of condensed tannins of sorghum genotypes. AACC Annual Meeting Abstracts. URL: /http:// www.aaccnet.org/meetings/2004/ abstracts/a04ma386.htmS.

Dykes, L., Rooney, L.W., Waniska, R.D., Rooney, W.L. (2005). Phenolic compounds and antioxidant activity of sorghum grains of varying genotypes. Journal of Agricultural and Food Chemistry 53, 6813–6818.

Earp, C.F., Akingbala, J.O., Ring, S.H., Rooney, L.W. (1981). Evaluation of several methods to determine tannins in sorghums with varying kernel characteristics. Cereal Chemistry 58, 234–238.

Earp, C.F., Doherty, C.A., Rooney, L.W. (1983). Fluorescence microscopy of the pericarp, aleurone layer and endosperm cell walls of three sorghum cultivars. Cereal Chemistry 60, 408–410.

Earp, C.F., McDonough, C.M., Awika, J., Rooney, L.W. (2004a). Testa development in the caryopsis of Sorghum bicolour (L.) Moench. Journal of Cereal Science 39, 303–311.

Earp, C.F., McDonough, C.M., Rooney, L.W. (2004b). Microscopy of pericarp development in the caryopsis of Sorghum bicolour (L) Moench. Journal of Cereal Science 39, 21–27.

Earp, C.F., Rooney, L.W. (1982). Scanning electron microscopy of the pericarp and testa of several sorghum varieties. Food Microstructure 1, 125–134.

Eidi, A., Eidi, M., and Darzi, R. (2009). Antidiabetic effect of Olea europaea L. in normal and diabetic rats. Phytotherapy Research, 23(3), 347–350.

Elkin, R.G., Featherston, W.R., Rogler, J.C. (1978b). Investigation of leg abnormalities in chicks consuming high tannin sorghum grain diets. Poultry Science 57, 757–762.

Elkin, R.G., Freed, M.B., Hamaker, B.R., Zhang, Y., Parsons, C.M. (1996). Condensed tannins are only partially responsible for variations in nutrient digestibilities of sorghum grain cultivars. Journal of Agricultural and Food Chemistry 44, 848–853.

Elkin, R.G., Rogler, J.C., Featherston, W.R. (1978a). Influence of sorghum grain tannins on methionine utilization in chicks. Poultry Science 57, 704–710.

Eller, M.S., Ostrom, K., Gilchrest, B.A. (1996). DNA damage enhances melanogenesis. Proceedings of the National Academy of Sciences of the United States of America 93, 1087–1092.

Ellis, E.B. (1972). Morphological characteristics in relation to seed deterioration in sorghum. M.S. Thesis, Texas A&M University, College Station, TX.

Emmons, C. L., and Peterson, D. M. (1999). Antioxidant activity and phenolic contents of oat groats and hulls. Cereal Chemistry, 76(6), 902–906.

Fang, N., Yu, S., Badger, T. (2003). Characterization of triterpene alcohol and sterol ferulates in rice bran using LC-MS/MS. Journal of Agricultural and Food Chemistry 51, 3260–3267.

Featherson, W.R., Rogler, J.C., 1975. Influence of tannin on the utilization of sorghum grain by rats and chickens. Nutrition Reports International 11, 491–497.

Fenster, C. (2003). White food sorghum in the American diet. In: US Grains Council 43rd Board of Delegates' Meeting, July 2003, Minneapolis, MN. Available from <http://www. grains.org/ news/ latest_news/43_delegates_meeting_updates.html>.

Flavonoids extracted from Fonio millet (Digitaria exilis) reveal potent antithyroid properties. Nutrition 12, 100–106.

Foo, L.Y., Lu, Y., Howell, A.B., Vorsa, N. (2000). A-type proanthocyanidin trimers from cranberry that inhibit adherence of uropathogenic P-fimbriated Escherichia coli. Journal of Natural Products 63, 1225–1228.

Fossati, P., Prgncipe, L., and Berti, G. (1980). Use of 3, 5-dichloro-2-hydroxybenzenesulfonic acid/4-aminophenazone chromogenic system in direct enzymic assay of uric acid in serum and urine. Clinical Chemistry, 26, 227–231.

Fraga, V., Menendez, R., Amor, A.M., Gonzalez, R.M., Jimenez, S., Mas, R. (1997). Effects of policosanol on in vitro and in vivo rat liver microsomal lipid peroxidation. Archives of Medical Research 28, 355–360.

Francis, F. J. (1989). Food colourants: Anthocyanins. Critical Reviews in Food Science and Nutrition, 28, 273–314.

Fulecki, T., and Francis, F. J. (1968). Quantitative methods for anthocyanins. 1. Extraction and determination of total anthocyanin in cranberries. Journal of Food Science, 33, 72–77.

Fuleki, T., Francis, F.J. (1968). Quantitative methods for anthocyanins. 2. Determination of total anthocyanin and degradation index for cranberry juice. Journal of Food Science 33, 78–83.

Garcia-Viguera, C., Zafrilla, P., and Tomas-Barberan, F. A. (1998). The use of acetone as an extraction solvent for anthocyanins from strawberry fruit. Phytochemical Analysis, 9, 247–277.

Glick, A., Joslyn, M.A. (1970b). Effect of tannic acid and related compounds on the absorption and utilization of proteins in the rat. Journal of Nutrition 100, 516–520.

Glick, Z., Joslyn, M.A. (1970a). Food intake depression and other metabolic effects of tannic acid in the rat. Journal of Nutrition 100, 509–515.

Gomez-Cordoves, C., Bartolome, B., Vieira, W., Virador, V.M. (2001). Effects of wine phenolics and sorghum tannins on tyrosinase activity and growth of melanoma cells. Journal of Agricultural and Food Chemistry 49, 1620–1624.

Gordon, L. A. (2001). Utilization of sorghum brans and barley flour in bread. M.S. Thesis, Texas AandM University: College Station, TX.

Gouni-Berthold, I., Berthold, H.K. (2002). Policosanol: clinical pharmacology and therapeutic significance of a new lipid-lowering agent. American Heart Journal 143, 356–365.

Gous, F. (1989). Tannins and phenols in black sorghum. Ph.D. Dissertation, Texas AandM University: College Station, TX.

Grimmer, H.R., Parbhoo, V., McGarth, R.M. (1992). Antimutagenicity of polyphenol-rich fractions from Sorghum bicolour grain. Journal of Agricultural and Food Chemistry 59, 251–256.

Grinberg, L.N., Newmark, H., Kitrossky, N., Rahamin, E., Chevion, M., Rachmilewitz, E.A. (1997). Protective effect of tea polyphenols against oxidative damage to red blood cell. Biochimica et Biophysica Acta 1201, 284–288.

Gu, L., Kelm, M. A., Hammerstone, J. F., Beecher, G., Holden, J., and Haytowitz, D. (2004). Concentration of proanthocyanidins in common foods and estimations of normal consumptions. The Journal of Applied Nutrition, 134, 613–617.

Gu, L., Kelm, M., Hammerstone, J.F., Beecher, G., Cunnigham, D., Vannozzi, S., Prior, L. (2002). Fractionation of polymeric procyandins from lowbush blueberry and quantifiation of procyanidins in selected foods with an optimized normal-phase HPLC-MS fluorescent detection method. Journal of Agricultural and Food Chemistry 50, 4852–4860.

Gu, L., Kelm, M.A., Hammerstone, J.F., Beecher, G., Holden, J., Haytowitz, D., Prior, R.L. (2003). Screening of foods containing proanthocyanidins and their structural characterization using LC–MS/MS and thiolytic degradation. Journal of Agricultural and Food Chemistry 51, 7513–7521.

Gujer, R., Magnolato, D., Self, R. (1986). Glucosylated flavonoids and other phenolic compounds from sorghum. Phytochemistry 25, 1431–1436.

Gupta, R.K., Haslam, E. (1978). Plant proanthocyanidins. 5. Sorghum polyphenols. Journal of the Chemical Society, Perkin Transactions 1 4 (8), 892–896.

Haber, B. (2002). Carob fiber benefits and applications. Cereal Foods World, 47(8), 365–373.

Hagerman, A.E., Butler, L.G. (1978). Protein precipitation method for the quantitative determination of tannins. Journal of Agricultural and Food Chemistry 26, 809–812.

Hagerman, A.E., Butler, L.G. (1980). Condensed tannin purification and characterization of tannin-associated proteins. Journal of Agricultural and Food Chemistry 28, 947–952.

Hagerman, A.E., Butler, L.G. (1981). The specificity of proanthocyanidin–protein interactions. Journal of Biological Chemistry 256, 4494–4497.

Hagerman, A.E., Riedl, K.M., Jones, G.A., Sovik, K.N., Ritchard, N.T., Hartzfeld, P.W., Riechel, T.L. (1998). High molecular weight plant polyphenolics (tannins) as biological antioxidants. Journal of Agricultural and Food Chemistry 46, 1887–1892.

Hagerman, A.E., Zhao, Y., Johnson, S. (1997). Methods for determination of condensed and hydrolysable tannins. In: Shahidi, F. (Ed.), Antinutrients and Phytochemicals in Food. American Chemical Society, Washington, DC, pp. 209–222.

Hahn, D. H. (1984). Phenols of sorghum and maize: The effect of genotype and alkali processing. Ph.D. Dissertation. Texas A and M

Hahn, D.H., Faubion, J.M., Rooney, L.W. (1983). Sorghum phenolic acids, their high performance liquid chromatography separation and their relation to fungal resistance. Cereal Chemistry 60, 255–259.

Hahn, D.H., Rooney, L.W. (1986). Effect of genotype on tannins and phenols of sorghum. Cereal Chemistry 63, 4–8.

Hahn, D.H., Rooney, L.W., Earp, C.F. (1984). Tannins and phenols of sorghum. Cereal Foods World 29, 776–779.

Hammerstone, J.F., Lazarus, S.A., Mitchel, A.E., Rucker, R., Schmitz, H.H. (1999). Identification of procyanidins in cocoa (Theobroma cocoa) and chocolate using high-performance liquid chromatography/mass spectrophotometry. Journal of Agricultural and Food Chemistry 47, 490–496.

Hammerstone, J.F., Lazarus, S.A., Schmitz, H.H. (2000). Procyanidin content and variation in some commonly consumed foods. Journal of Nutrition 130, 2086s–2092s.

Harold H R, Reeves R D, Reeves R D, Bulze M S, Guthrie D. W. (1985). Effect of dietary fibre in IDDM insulin requirement and serum lipids. J.Am.Dietet. Assoc. 85:1455-60.

Haslam, E. (1974). Polyphenol–protein interactions. Biochemical Journal 139, 285–288.

Hegde, P.S., Chandra, T.S. (2005). ESR spectroscopic study reveals higher free radical quenching potential in kodo millet (*Paspalum scrobiculatum*) compared to other millets. Food Chemistry 92, 177–182.

Heinonen, I.M., Meyer, A.S., Frankel, E.N. (1998). Antioxidant activity of berry phenolics on human low-density lipoprotein and liposome oxidation. Journal of Agricultural and Food Chemistry 46, 4107–4112.

Helsper, J.P.F.G., Kolodziej, H., Hoogendijk, J.M., Van Norel, A. (1993). Characterization and trypsin inhibitor activity of proanthocyanidins from Vicia faba. Phytochemistry 34, 1255–1260.

Higdon, J.V., Frei, B. (2003). Tea catechins and polyphenols: health effects, metabolism, and antioxidant functions. Critical Reviews in Food Science and Nutrition 43, 89–143.

Hill, J.O., Wyatt, H.R., Reed, G.W., Peters, J.C. (2003). Obesity and the environment: where do we go from here? Science 299, 853–855.

Hilu, K.W., De Wet, J.M.J., Seigler, D. (1978). Flavonoid patterns and systematics in Eleusine. Biochemical Systematics and Ecology 6, 247–249.

Hollnbeck C B, Coulston A M, Donner C C, Williams R A, Reaven G M. (1985). The effect of variations in per cent of naturally occuring complex and simple carbohydrates on plasma glucose and insulin response in individuals with non-insulin dependent diabetes mellitus. Diabetes 34:151-155.

Holman, R. R., and Turner, R. C. (1991). Oral agents and insulin in the treatment of NIDDM. In J., and G. (Eds.), Textbook of diabetes (pp. 467–469). Oxford: Blackwell.

http://www.cdc.gov/nchs/nhanes/nhanes2005-2006/nhanes05_06.htm

http://www.cnpp.usda.gov/publications/dietaryguidelines/2010/dgac/report/d-5-carbohydrates.pdf

http://www.fda.gov/Food/GuidanceComplianceRegulatoryInformation/GuidanceDocuments/FoodLabelingNutrition/FoodLabelingGuide/ucm064919.htm

Hultcrantz, R., Glaumann, H., Lindberg, G., and Nilsson, L.H. (1986). Liver investigation in asymptomatic patients with moderately elevated activities of serum aminotransferases. Scandinavian Journal of Gastroenterology, 21, 109.

Hwang, K.T., Cuppett, S.L., Weller, C.L., Hanna, M.A. (2002). Properties, composition, and analysis of grain sorghum wax. Journal of the American Oil Chemists Society 79, 521–526.

Ill-Min Chung, Eun-Hye Kim, Min-A Yeo, Sun-Jin Kim, Myong–Cheol Seo and Hyung-In Moon. (2011). Antidiabetic effects of three Korean sorghum phenolic extracts in normal and streptozotocin-induced diabetic rats. Food Research International 44: 127–132.

ISO, (1988). Sorghum. Determination of tannin content. International Organization for Standardization, ISO 9648:1988.

Jacob, J.P., Mitaru, B.N., Mbugua, P.N., Blair, R. (1996). The effects of substituting Kenyan Serena sorghum for maize in broiler starter diets with different dietary crude protein and methionine levels. Animal Feed Science and Technology 61, 27–39.

Jambubathan, R., Mertz, E.T. (1973). Relationship between tannin levels, rat growth, and distribution of proteins in sorghum. Journal of Agricultural and Food Chemistry 21, 692–696.

Jambunathan, R., Kherdekar, M.S., Bandyopadhyay, R. (1990). Flavan-4-ols concentration in mold-susceptible and mold-resistant sorghum at different stages of grain development. Journal of Agricultural and Food Chemistry 38, 545–548.

Jambunathan, R., Kherdekar, M.S., Vaidya, P. (1991). Ergosterol concentration in mold-susceptible and mold-resistant sorghum at different stages of grain development and its relationship to flavan-4-ols. Journal of Agricultural and Food Chemistry 39, 1866–1870.

Jenkins D J A, Wolever T M S, Bacon S, Nineham R, Leeds R, Rowden R, Love M, Hockaday T D Rr. (1980). Diabetic diets: high carbohydrate combined with high-fibre. Am. J.Clin. Nutr. 33:1729-1733.

Jenkins D J A. (1981). Slow release carbohydrates and treatment of diabetes. Proc. Nutr. Sot. India. 40: 227-235.

Ji Heon Park, Sun Hee Lee, Ill-Min Chung and Yongsoon Park. (2012). Sorghum extract exerts an anti-diabetic effect by improving insulin sensitivity via PPAR-γ in mice fed a high-fat diet. Nutr Res Pract. 2012 Aug; 6(4): 322–327.

Jonker, D., van der Hoek, G.D., Glatz, C.H., Homan, C., Posthumus, M.A., Katan, M.B. (1985). Combined determination of free, esterified and glycosylated plant sterols in foods. Nutrition Reports International 32, 943–951.

Joseph M. Awika and Lloyd W. Rooney. (2004). Sorghum phytochemicals and their potential impact on human health. Phytochemistry. 65: 1199–1221.

Ju-Sung Kim, Tae Kyung Hyun and Myong-Jo Kim. (2011). The inhibitory effects of ethanol extracts from sorghum, foxtail millet and proso millet on a-glucosidase and a-amylase activities. Food Chemistry 124: 1647–1651.

Kallithraka, S., Garcia-Viguera, C., Bridle, P., and Bakker, J. (1995). Survey of solvents for the extraction of grape seed phenolics. Phytochemical Analysis, 6, 265–267.

Kaluza, W. Z., McGrath, R. M., Roberts, T. C., and Schroder, H. H. (1980). Separation of phenolics of Sorghum bicolour (L.) Moench grain. Journal of Agricultural and Food Chemistry, 28(6), 1191–1196.

Kamath, V. G., Chandrashekar, A., and Rajini, P. S. (2004). Antiradical properties of sorghum (Sorghum bicolour L. Moench) flour extracts. Journal of Cereal Science, 40, 283–288.

Kambal, A.E., Bate-Smith, E.C. (1976). A genetic and biochemical study on pericarp pigments in a cross between two cultivars of grain sorghum, Sorghum bicolour. Heredity 37, 413–416.

Kamei, H., Kojima, T., Hasegawa, M., Koide, T., Umeda, T., Yukawa, T., and Terabe, K. (1995). Suppression of tumor cell growth by anthocyanins in vitro. Cancer Investigations, 13(6), 590–594.

Kamini Devi. (1972). Comparative observations on blood sugar level in normal adults following common cereal and millet diets. M.Sc. (Foods & Nutrition) thesis submitted to A.P.A.U., Hyderabad.

Karaivanova, M., Drenska, D., and Ovcharov, R. (1990). A modification of the toxic effects of platinum complexes with anthocyanins. Eksperimetnalna Meditsna I Morfologiia, 29(2), 19–24.

Keli, S.O., Hertog, M.G., Fesken, E.J., Kromhout, D. (1996). Dietary flavonoids, antioxidant vitamins, and incidences of stroke: the Zutphen study. Archives of Internal Medicine 156, 637–642.

Kennedy, J.A., Hayasaka, Y., Vidal, S., Waters, E.J., Jones, G.P. (2001). Composition of grape skin proanthocyanidins at different stages of berry development. Journal of Agricultural and Food Chemistry 49, 5348–5355.

Kiehm T G, Anderson J W, Ward K. (1976). Beneficial effects of a highcarbohydrate high-fibre diet in hyperglycemic men. Am. J. Clin. Nutr. 1976: 29:895-899.

Kim JS, Hyun TK, and Kim MJ. (2011). The inhibitory effects of ethanol extracts from sorghum, foxtail millet and proso millet on α-glucosidase and α-amylase activities. Food Chem. 124:1647–1651.

Kim, E. H., Kim, S. H., Chung, J. I., Chi, H. Y., Kim, J. A., and Chung, I. M. (2006). Analysis of phenolic compounds and isoflavones in soybean seeds (*Glycine max* (L.) Merill) and sprouts grown under different conditions. European Food Research and Technology, 222, 201–208.

King, D., Fan, M.Z., Ejeta, G., Asem, E.K., Adeola, O. (2000). The effects of tannins on nutrient utilization in the White Pekin duck. British Poultry Science 41, 629–630.

Klopfenstein, C.F., Varriano-Marston, E., Hoseney, R.C. (1981). Cholesterol-lowering effect of sorghum diet in guinea pigs. Nutrition Reports International 24, 621–627.

Kondos, A.C., Foale, M.A. (1983). Comparison of the nutritional value of low and medium tannin sorghum grains for pigs. Animal Feed Science and Technology 8, 85–90.

Kroon, P.A., Faulds, C.B., Ryden, P., Robertson, J.A., Williamson, G. (1997). Release of covalently bound ferulic acid from fiber in human colon. Journal of Agricultural and Food Chemistry 45, 661–667.

Krueger, C.G., Vestling, M.A. and Reed, J.D. (2003). Matrix-assisted laser desorption/ionization time-of-flight mass spectrometry of heteropolyflavan-3-ols and glucosylated heteropolyflavans in sorghum (*Sorghum bicolour* (L.) Moench). Journal of Agricultural and Food Chemistry 51, 538–543.

Kushi, L.H., Meyer, K.A. and Jacobs, D.R. (1999). Cereals, legumes, and chronic disease risk reduction: evidence from epidemiologic studies. American Journal of Clinical Nutrition 70 (Suppl.), 451S–458S.

Lacobucci, G.A. and Sweeny, J.G. (1983). The chemistry of anthocyanins, anthocyanidins and related flavylium salts. Tetrahedron, 39(19), 3005–3038.

Lakshmi KB, and Vimala V. (1996). Hypoglycemic effect of selected sorghum recipes. Nutr Res. 16: 1651–1658.

Lapidot, T., Harel, S., Akiri, B., Granit, R. and Kanner, J. (1999). pH dependent forms of red wine anthocyanins as antioxidants. Journal of Agricultural and Food Chemistry 47, 67–70.

Lea, A.G.H., Bridle, P., Timberlake, C.F. and Singleton, V.L. (1979). The procyanidins of white grapes and wines. American Journal of Enology and Viticulture.

Lee SH, Chung IM, Cha YS and Park Y. (2010). Millet consumption decreased serum concentration of triglyceride and C-reactive protein but not oxidative status in hyperlipidemic rats. Nutr Res. 30:290–296.

Lee, K.H., Tagahara, K., Suzuki, H., Wu, R.-Y., Haruna, M. and Hall, I.H. (1981). Antitumor agents. 49. Tricin, kaempferol-3-O-b-D-glucopyranoside and (+)-nortrachelogenin, antileukemic principles from Wikstroemia indica. Journal of Natural Products 44, 530–535.

Lee, S.M. and Pan, B.S. (2003). Effect of dietary sorghum distillery residue on hematological characteristics of cultured grey mullet (Mugil cephalus) - an animal model for prescreening antioxidant and blood thinning activities. Journal of Food Biochemistry 27, 1–18.

Lietti, A., Cristoni, A. and Picci, M. (1976). Studies of Vaccinium myrtillus anthocyanosides. I. Vasoprotective and anti-inflammatory activity. Arzneimittel-Forschung, 26(5), 829–832.

Lin, B.B., Chen, H.L. and Huang, P.C. (1986). Effects of instant Pauchong tea, catechin, and caffeine on serum-cholesterol and serum lowdensity-lipoprotein in mice. Nutrition Reports International 34, 821–829.

Linda Dykes and Lloyd W. Rooney. (2006). Sorghum and millet phenols and antioxidants. Journal of Cereal Science. 44: 236–251.

Lizardo, R., Peiniau, J. and Aumaitre, A. (1995). Effect of sorghum on performance, digestibility of dietary-components and activities of pancreatic and intestinal enzymes in the weaned piglet. Animal Feed Science and Technology 56, 67–82.

Lo, S.C.C., De Verdier, K. and Nicholson, R.L. (1999). Accumulation of 3-deoxyanthocyanidin phytoalexins and resistance to Colletotrichum sublineolum in sorghum. Physiological and Molecular Plant Pathology 55, 263–273.

Lotito, S.B., Actis-Goretta, L., Renart, M.L., Caligiuri, M., Rein, D., Schmitz, H.H., Steinberg, F.M., Keen, C.L. and Fraga, C.G. (2000). Influence of oligomer chain length on antioxidant activity of procyanidins. Biochemical and Biophysical Research Communications 276, 945–951.

Lu, Y. and Foo, L.Y. (2001). Unusual anthocyanin reactions with acetone leading to pyranoanthocyanin formation. Tetrahedron Letters, 42, 1371–1373.

Madinov, I.V., Balabolkin, M.I., Markov, D.S. and Markova, T.N. (2000). Main causes of hyperuricemia in diabetes mellitus. Terapevticheskii Arkhiv, 72, 55–58.

Maillard, M.N. and Berset, C. (1995). Evolution of antioxidant activity during kilning – role of insoluble bound phenolic-acids of barley and malt. Journal of Agricultural and Food Chemistry 43, 1789–1793.

Makkar, H.P.S. (1989). Protein precipitation methods for quantitation of tannins: a review. Journal of Agricultural and Food Chemistry 37, 1197–1202.

Maron DJ, Lu GP, Cai NS, Wu ZG, Li YH, Chen H, Zhu JQ, Jin XJ, Wouters BC and Zhao J. (2003). Cholesterol-lowering effect of a theaflavin-enriched green tea extract: a randomized controlled trial. Arch Intern Med. 163:1448–1453.

Matsumoto, H., Inaba, H., Kishi, M., Tominaga, S., Hirayama, M. and Tsuda, T. (2001). Orally administered delphinidin-3-rutinoside and cyanidin-3-rutinoside are directly absorbed in rats and humans and appear in blood as the intact forms. Journal of Agricultural and Food Chemistry 49, 1546–1551.

Maxson, E.D., Rooney, L.W. (1972). Evaluation of methods for tannin analysis in sorghum grain. Cereal Chemistry 49, 719–729.

Mazza, G. and Brouillard, R. (1987). Colour stability and structural transformations of cyaniding 3,5-diglucoside and four 3-deoxyanthocyanins in aqueous solutions. Journal of Agricultural and Food Chemistry 35, 422–426.

McCarthy, M.F. (2002). Policosanol safely down-regulates HMG-CoA reductase – potential as a component of the Esselstyn regimen. Medical Hypotheses 59, 268–279.

McDonough, C.M. and Rooney, L.W. (2000). The millets. In: Kulp, K., Ponte, Jr., J.G. (Eds.), Handbook of Cereal Science and Technology. Marcel Dekker, Inc., New York, pp. 177–201.

McDonough, C.M., Rooney, L.W. and Earp, C.F. (1986). Structural characteristics of *Eleusine coracana* (finger millet) using scanning electron and fluorescence microscopy. Food Microstructure 5, 247–256.

Melake-Berhan, A., Butler, L.G., Ejeta, G. and Menkir, A. (1996). Grain mold resistance and polyphenol accumulation in sorghum. Journal of Agricultural and Food Chemistry 44, 2428–2434.

Menkir, A., Ejeta, G., Butler, L. and Melakeberhan, A. (1996). Physical and chemical kernel properties associated with resistance to grain mold in sorghum. Cereal Chemistry 73, 613–617.

Milbury, P.E., Cao, G., Prior, R.L. and Blumberg, J. (2002). Bioavailability of elderberry anthocyanins. Mechanisms of Aging and Development 123, 997–1006.

Misra, K. and Seshadri, T.R. (1967). Chemical components of sorghum durra glumes. Indian Journal of Chemistry 5, 409.

Mitre-Dieste, C. M., Gordon, L. A., Awika, J., Suhendro, E.L. and Rooney, L.W. (2000). Cookies made with sorghum brans high in phenols and catechins. AACC Annual Meeting Abstracts. Available: http://www.aaccnet.org/meetings/2000/ abstracts/aOOma203. htm.

Moreau, R.A., Powell, M.J. and Hicks, K.B. (1996). Extraction and quantitative analysis of oil from commercial corn. Journal of Agricultural and Food Chemistry 44, 2149–2154.

Morton, J.F. (1970). Tentative correlations of plant usage and esophageal cancer zones. Economic Botany 24, 217–226.

Morton, J.F. (1972). Further association of plant tannins and human cancer. Quarterly Journal of Crude Drug Research 12, 1829–1841.

Moss, D.W. and Henderson, A.R. (1999). Clinical enzymology. In C.A. and E.R. (Eds.), Tietz textbook of clinical chemistry (pp. 617–721). (3 rd edn). Philadelphia: WB Saunders.

Moyer, A.R., Humer, K.E., Finn, C.E., Frei, B. and Wrolstad, R.E. (2002). Anthocyanins, phenolics, and antioxidant capacity in diverse small fruits: Vaccinium, Rubus, and Ribers. Journal of Agricultural and Food Chemistry 50, 519–525.

Muriu, J.I., Njoka-Njiru, E.N., Tuitoek, J.K. and Nanua, J.N. (2002). Evaluation of sorghum (*Sorghum bicolour*) as replacement for maize in the diet of growing rabbits (*Oryctolagus cuniculus*). Asian-Australian Journal of Animal Science 15, 565–569.

Murthy, D.S. and Kumar, K. A. (1995). Traditional uses of sorghum and millets. In D. A. V. Dendy (Ed.), Sorghum and millets: Chemistry and technology (pp. 185–221). St Paul, MN, USA: American Association of Cereal Chemistry.

Naczk, M. and Shahidi, F. (1997). Nutritional implications of canola condensed tannins. Antinutrients and Chemicals in Food: ACS Symposium Series 662, 186–208.

Naczk, M. and Shahidi, F. (2004). Extraction and analysis of phenolics in food. Journal of Chromatography A 1054, 95–111.

Nguz, K., Van Gaver, D. and Huyghebaert, A. (1998). In nitro inhibition of digestive enzymes by sorghum condensed tannins (*Sorghum bicolour* L. (Moench)). Sciences des Aliments 18, 507–514.

Nip, W.K. and Burns, E. E. (1969). Pigment characterization in grain sorghum. I. Red varieties. Cereal Chemistry, 46, 490–495.

Nip, W.K. and Burns, E. E. (1971). Pigment characterization in grain sorghum. II. White varieties. Cereal Chemistry, 48, 74–80.

Nordkvist, E., Salomonsson, A.C. and Aman, P. (1984). Distribution of insoluble bound phenolic acids in barley grain. Journal of the Science of Food and Agriculture 35, 657–661.

Nyachoti, CM., Atkinson, J.L. and Leeson, S. (1997). Sorghum tannins: a review. Poultry Science 53, 5–21.

O'Dea K, Nestel P.J. and Antonoff L. B. (1980). Physical factors influencing post-prandial glucose and insulin responses to starch. Am. J. Clin. Nutr. 33:760-765.

Osluzzi R and Crivellent G. (1978). Effect of high fiber diet on blood qlucose and plasma insulin in diabetic subjects. Diabetelogia 15:259.

Osman, M.A. (2004). Changes in sorghum enzyme inhibitors, phytic acid, tannins and *in vitro* protein digestibility occurring during Khamir (local bread) fermentation. Food Chemistry 88, 129–134.

Ostlund, R.E. (2002). Phytosterols in human nutrition. Annual Reviews of Nutrition 22, 533–549.

Oterdoom, H.J. (1985). Tannin, sorghum, and oesophageal cancer. Lancet 2 (8450), 330.

Ozsoy-Sacan, O., Karabulut-Bulan, O., Bolkent, S., Yanardag, R., and Ozgey, Y. (2004). Effects of chard (*Beta vulgaris* L. var *cicla*) on the liver of the diabetic rats: a morphological and biochemical study. Bioscience, Biotechnology, and Biochemistry, 68, 1640–1648.

Pace-Asciak, C.R., Rounova, O., Hahn, S.E., Diamandis, E.P. and Goldberg, D.M. (1996). Wine and grape juice as modulators of platelet aggregation in healthy human subjects. Clinica Chimica Acta 246, 163–182.

Pale, E., Kouda-Bonafos, M., Mouhoussine, N., Vanhaelen, M., Vanhaelen-Fastre, R. and Ottinger, R. (1997). 7-O-methylapigeninidin, an anthocyanidin from Sorghum Caudatum. Phytochemistry 45, 1091–1092.

Parbhoo, V., Grimmer, H.R., Cameron-Clarke, A. and McGarth, R.M. (1995). Induction of cytochrome P-450 in rat liver by a polyphenolrich extract from a bird-resistant sorghum grain. Journal of the Science of Food and Agriculture 69, 247–252.

Paulis, J.W. and Wall, J.S. (1979). Distribution and electrophoretic properties of alcoholsoluble proteins in normal and high-lysine sorghums. Cereal Chemistry, 56, 20–23.

Pepping, J. (2003). Policosanol. American Journal of Health-System Pharmacy 60, 1112–1114.

Perez-Gonzalez, A. (2005). Specialty sorghums in direct-expansion extrusion. M.S. Thesis, Texas AandM University, College Station, TX.

Pietta, P.G., Gardana, C. and Mauri, P.L. (1997). Identification of Ginkgo biloba flavonol metabolites after oral administration to humans. Journal of Chromatography 693, 249–255.

Piironen, V., Toivo, J. and Lampi, A.M. (2002). Plant sterols in cereals and cereal products. Cereal Chemistry 79, 148–154.

Pinto, A. P., Mota, A. M., de Varennes, A. and Pinto, F. C. (2004). Influence of organic matter on the uptake of cadmium, zinc, copper and iron by sorghum plants. Science of the total Environment. 326(1-3): 239-247.

Porter, L.J., Hrstich, L.N. and Chan, B.G. (1986). The conversion of procyanidins and prodelphinidins to cyanidin and delphinidin. Phytochemistry 25, 223–230.

Price, M.L. and Butler, L.G. (1977). Rapid visual estimation and spectrophotometric determination of tannin content of sorghum grain. Journal of Agricultural and Food Chemistry 25, 1268–1273.

Price, M.L., Butler, L.G., Rogler, J.C. and Featherston, W.R. (1979). Overcoming the nutritionally harmful effects of tannin in sorghum grain by treatment with inexpensive chemicals. Journal of Agricultural and Food Chemistry 27, 441–445.

Price, M.L., Van Scoyoc, S. and Butler, L.G. (1978). A critical evaluation of the vanillin reaction as an assay for tannin in sorghum grain. Journal of Agricultural and Food Chemistry 26, 1214–1218.

Prior, R.L., Cao, G., Martin, A., Sofic, E., McEwens, J., O'Brien, C., Lischner, N., Ehlenfeldt, M., Kalt, W., Krewer, G. and Mainland, C.M. (1998). Antioxidant capacity as influenced by total phenolic and anthocyanin content, maturity, and variety of Vaccinium species. Journal of Agricultural and Food Chemistry 46, 2686–2693.

Prior, R.L. and Gu, L. (2005). Occurrence and biological significance of proanthocyanidins in the American diet. Phytochemistry 66, 2264–2280.

Putman, L.J. and Butler, L.G. (1989). Separation of high molecular weight sorghum procyanidins by high performance liquid chromatography. Journal of Agricultural and Food Chemistry 37, 943–946.

Quettier-Deleu, C., Gressier, B., Vasseur, J., Dine, T., Brunet, C., Luyckx, M., Cazin, M., Cazin, C., Bailleul, F. and Trotin, F. (2000). Phenolic compounds and antioxidant activities of buckwheat (*Fagopyrum esculentum* Moench) hulls and flour. Journal of Ethnopharmacology, 72(1), 35–42.

Ramachandra, G., Virupaksha, T.K. and Shadaksharaswamy, M. (1977). Relationship between tannin levels and in vitro protein digestibility in finger millet (*Eleusine coracana* Gaertn.). Journal of Agricultural and Food Chemistry 25, 1101–1104.

Ray TK, Mansell KM, Knight LC, Malmud LS, Owen OE and Boden G. (1983). Long-term effects of dietary fiber on glucose tolerance and gastric emptying in noninsulin-dependent diabetic patients. Am J Clin Nutr. 37:376–381.

Rayudu, G.V.N., Kakirvel, R., Vohra, P. and Kratzer, F.H. (1970). Toxicity of tannic acid and its metabolites for chickens. Poultry Science 49, 957–960.

Reichert, R.D. (1979). The pH-sensitive pigments in pearl millet. Cereal Chemistry 56, 291–294.

Reichert, R.D., Fleming, S.E. and Schwab, D.J. (1980). Tannin deactivation and nutritional improvement of sorghum by anaerobic storage of H2O-, HCl-, or NaOH-treated grain. Journal of Agricultural and Food Chemistry 28, 824–829.

Remy, S., Fulcrand, H., Labarbe, B., Cheynier, V. and Moutonet, M. (2000). First confirmation in red wine of products resulting from direct anthocyanin–tannin reactions. Journal of the Science of Food and Agriculture 80, 745–751.

Renaud, S. and De Lorgeril, M. (1992). Wine, alcohol, platelets, and the French paradox for coronary heart disease. Lancet 339, 1523–1526.

Riedl, K.M. and Hagerman, A.E. (2001). Tannin–protein complexes as radical scavengers and radical sinks. Journal of Agricultural and Food Chemistry 49, 4917–4923.

Rifai, N., Bachorik, P.S. and Albers, J.J. (1999). Lipids, lipoproteins and apolipoproteins. In C.A. and E.R. (Eds.), Tietz textbook of clinical chemistry (pp. 809–861). (3 rd edn). Philadelphia: WB Saunders.

Rigaud, J., Escribano-Bailon, M.T., Prieur, C., Souquet, J.M. and Cheynier, V. (1993). Normal-phase high-performance liquid chromatographic separation of procyanidins from cacao beans and grape seeds. Journal of Chromatography A 654, 255–260.

Ring, A.S. (1984). Phenolic compounds in sorghum from boot stage to post harvest maturity by Folin-Ciocalteau and high-pressure liquid chromatography methods. M.S. Thesis, Texas AandM University, College Station, TX.

Rios, L.Y., Bennett, R.N., Lazarus, S.A., Remesy, C., Scalbert, A. and Williamson, G. (2002). Cocoa procyanidins are stable during gastric transit in humans. American Journal of Clinical Nutrition 76, 1106–1110.

Rogers, E.J., Rice, S.M., Nicolosi, R.J., Carpenter, D.R., McClelland, C.A. and Romanczyk, L.J. (1993). Identification and quantification of y-oryzanol components and simultaneous assessment of tocols in rice bran oil. Journal of the American Oil Chemists Society 70, 301–307.

Rohr, G.E., Meier, B. and Sticher, O. (2000). Analysis of procyanidins. Studies in Natural Products Chemistry 21, 497–570.

Rooney, L.W. and Waniska, R. D. (2000). Sorghum food and industrial utilization. In C. W. Smith, and R. W. Frederiksen (Eds.), Sorghum, origin, history, technology and production (pp. 689–729). New York: Wiley.

Rooney, L.W. (2001). Food and nutritional quality of sorghum and millet. INTSORMIL 2001 Annual Report, Project TAM-226, pp. 105–114. Available from: <http://intsormil. org/2001 anlrpt/2001TAM-226.pdf>.

Rooney, L.W. (2005). Ten myths about tannins in sorghums. International Sorghum and Millets Newsletter 46, 3–5.

Rooney, L.W., Miller, F.R. and Mertin, J.V. (Eds.), (1982). Proceedings of the International Symposium on Sorghum Grain Quality. ICRISAT, Patancheru, AP, India, pp. 143–162.

Rooney, L.W. and Sullins, R.D. (1977). The structure of sorghum and its relation to processing and nutritional value. In: Dendy, D.A.V. (Ed.), Proceedings of a Symposium on Sorghums and Millets for Human Food. Tropical Products Institute, London, pp. 91–109.

Rooney, L.W. and Waniska, R.D. (2000). Sorghum food and industrial utilization. In: Smith, C.W., Frederiksen, R.A. (Eds.), Sorghum: Origin, History, Technology, and Production. Wiley, New York, pp. 689–729.

Rooney, T.K., Rooney, L.W. and Lupton, J.R. (1992). Physiological characteristics of sorghum and millet brans in the rat model. Cereal Foods World 37, 782–786.

Ross, J.A. and Kasum, C.M. (2002). Dietary flavonoids: bioavailability, metabolic effects, and safety. Annual Reviews of Nutrition 22, 19–22.

Rostagno, H.S., Featherston, W.R. and Rogler, J.C. (1973). Studies on the nutritional value of sorghum grains with varying tannin contents for chicks. Poultry Science 52, 765–772.

Rudiger, C. (2003). The formulation of a nutraceutical bread mix using sorghum, barley, and flaxseed. M.S. Thesis, Texas AandM University, College Station, TX.

Ruzaidi A, Amin I, Nawalyah AG, Hamid M. and Faizul HA. (2005). The effect of Malaysian cocoa extract on glucose levels and lipid profiles in diabetic rats. J Ethnopharmacol. 98:55–60.

Ryu, S.N., Park, S.Z. and Ho, C.T. (1998). High performance liquid chromatographic determination of anthocyanin pigments in some varieties of black rice. Journal of Food and Drug Analysis 6, 729–736.

Santos-Buelga, C. and Scalbert, A. (2000). Proanthocyanidins and tanninlike compounds – nature, occurrence, dietary intake and effects on nutrition and health. Journal of the Science of Food and Agriculture 80, 1097–1117.

Sarni-Manchado, P., Cheynier, V. and Moutounet, M. (1999). Interactions of grape seed tannins with salivary proteins. Journal of Agricultural and Food Chemistry 47, 42–47.

Sato, Y., Nakatsuka, H., Watanabe, T., Hisamichi, S., Shimizu, H., Fujisaku, S., Ichinowatari, Y., Ida, Y., Suda, S., Kato, K. and Ikeda, M. (1989). Possiblc contribution of green tea drinking habits to the prevention of stroke. Tohoku Journal of Experimental Medicine 157, 337–343.

Scalbert, A., Morand, C., Manach, C. and Remesy, C. (2002). Absorption and metabolism of polyphenols in the gut and impact on health. Biomedicine and Pharmacotherapy 56: 276–282.

Schofield, P., Mbugua, D.M. and Pell, A.N. (2001). Analysis of condensed tannins: a review. Animal Feed Science and Technology 91, 21–40.

Scitz, L.M. (1977). Composition of sorghum grain wax. Cereal Foods World 22, 247.

Seitz, L.M. (2004). Effect of plant-type (purple vs. tan) and mold invasion on concentrations of 3-deoxyanthocyanidins in sorghum grain. AACC Annual Meeting Abstracts. URL: /http://www.aaccnet.org/meetings/2004/abstracts/a04ma384.htmS.

Sene, M., Dore, T. and Gallet, C. (2001). Relationship between biomass and phenolic production in grain sorghum grown under different conditions. Agronomy Journal, 93(1), 49–54.

Serna-Saldivar, S. and Rooney, L.W. (1995). Structure and chemistry of sorghum and millets. In D.A.V. (Ed.), Sorghum and millets: Chemistry and technology (pp. 69–124). (1st ed). St. Paul, MN: American Association of Cereal Chemists, Inc.

Sezik, E., Asla, M., Yesilada, E. and Ito, S. (2005). Hypoglycaemic activity of Gentiana olivieri and isolation of the active constituent through bioassay-directed fractionation techniques. Life Sciences, 76, 1223–1238.

Shahidi, F. and Naczk, M. (1995). Food Phenolics. Technomic Publishing Co., Inc., Lancaster, PA.

Shanaaz Saldana (1991). The effect of tannin and fibre on the in vitro protein digestibility of home scale processed sorghum products. M.Sc. (Foods & Nutrition) thesis submitted to the A.P. Agricultural University, Hyderabad (Unpublished).

Singh D.K., Banerjee S. and Porter TD. (2009). Green and black tea extracts inhibit HMG-CoA reductase and activate AMP kinase to decrease cholesterol synthesis in hepatoma cells. J Nutr Biochem. 20:816–822.

Singh, V., Moreau, R.A. and Hicks, K.B. (2003). Yield and phytosterol composition of oil extracted from grain sorghum and its wet-milled fractions. Cereal Chemistry 80 (2), 126–129.

Singleton, V.L. and Rossi, J.A. (1965). Colourimetry of total phenols with phosphomolybdic phosphotungstic acid reagents. American Journal of Enology and Viticulture 16, 144–158.

Sistino, J.J. (2003). Epidemiology of cardiovascular disease in the last decade: treatment options and implications for perfusion in the 21st century. Perfusion 18, 73–77.

Slavin, J.L., Jacobs, D. and Marquart, L. (2000). Grain processing and nutrition. Critical Reviews in Food Science and Nutrition 40, 309–326.

Spencer, J.P., Abd El Mohsen, M.M., Minihane, A.M. and Mathers, J.C. (2008). Biomarkers of the intake of dietary polyphenols: Strengths, limitations and application in nutrition research. The British Journal of Nutrition, 99(1), 12–22.

Spencer, J., Chaudry, F., Pannala, A., Srai, S., Debnam, E. and Rice-Evans, C. (2000). Decomposition of cocoa procyanidins in the gastric milieu. Biochemical and Biophysical Research Communications 272, 236–241.

Sripriya, G., Chandrasekharan, K., Murty, V.S. and Chandra, T.S. (1996). ESR spectroscopic studies on free radical quenching action of finger miller (Eleusine coracana). Food Chemistry 57, 537–540.

Stafford, H.A. (1965). Flavonoids and related phenolic compounds produced in the first internode of Sorghum vulgare preserved in darkness and in light. Plant Physiology, 40, 130–139.

Steadman, K.J., Burgoon, M.S., Lewis, B.A., Edwardson, E. and Obendorf, R.L. (2001). Minerals, phytic acid, tannin and rutin in buckwheat seed milling fractions. Journal of the Science of Food and Agriculture 81, 1094–1100.

St Leger, A.S., Cochrane, A.L. and Moore, F. (1979). Factors associated with cardiac mortality in developed countries with particular reference to consumption of wine. Lancet 1 (8124), 1017–1020.

Subba Rao, M.V.S.S.T. and Muralikrishna, G. (2001). Non-starch polysaccharides and bound phenolic acids from native and malted finger millet (Ragi, Eleusine coracana, Indaf-15). Food Chemistry 72, 187–192.

Subba Rao, M.V.S.S.T. and Muralikrishna, G. (2002). Evaluation of the antioxidant properties of free and bound phenolic acids from native and malted finger millet (Ragi, Eleusine coracana Indaf-15). Journal of Agricultural and Food Chemistry 50, 889–892.

Sweeny, J.G. and Iacobucci, G.A. (1981). Synthesis of anthocyanidins – III: total synthesis of apigeninidin and luteolinidin chlorides. Tetrahedron 37, 1481–1483.

Sweeny, J.G. and Iacobucci, G.A. (1983). Effect of substitution on the stability of 3-deoxyanthocyanidins in aqueous solutions. Journal of Agricultural and Food Chemistry 31, 531–533.

Tapiero, H., Tew, K.D., Nguyen, B.G. and Mathe, G. (2002). Polyphenols: do they play a role in the prevention of human pathologies? Biomedicine and Pharmacotherapy 56, 200–207.

Taskinen, M. R. (2002). Diabetic dyslipidemia. Atheroscler, S3, 47–51.

Tebib, K., Rouanet, J.M. and Besancon, P. (1997). Antioxidant effects of dietary polymeric grape seed tannins in tissue of rats fed a high cholesterol-vitamin E-deficient diet. Food Chemistry 59, 135–141.

Tedesco, I., Russo, M., Russo, P., Iacomino, G., Russo, G.L., Carraturo, F., Faruolo, C. and Moio, L., 2000. Antioxidant effect of red wine polyphenols on red blood cells. Journal of Nutritional Biochemistry 11, 114–119.

Thamer C, Machann J, Tschritter O, Haap M, Wietek B, Dahl D, Bachmann O, Fritsche A, Jacob S, Stumvoll M, Schick F. and Häring HU. (2002). Relationship between serum adiponectin concentration and intramyocellular lipid stores in humans. Horm Metab Res. 34:646–649.

Timberlake, C.F. and Bridle, P. (1980). Anthocyanins. In L. Walford (Ed.), Development of food colours – I (pp. 115–149). London: Applied Science.

Timberlake, C.F. and Bridle, P. (1980). Anthocyanins. In: Walford, J. (Ed.), Developments in Food Colour – 1. Applied Science, London, pp. 115–149.

Tipton, K.W., Floyd, E.H., Marshall, J.G. and McDevitt, J.B. (1970). Resistance of certain grain sorghum hybrids to bird damage in Louisiana. Agronomy Journal 62, 211–213.

Tomas, L. (1998a). Clinical laboratory diagnostics (pp. 208–214). (1st edn.). Frankfurt: TH-books Verlagsgesellschaft.

Tomas, L. (1998b). Clinical laboratory diagnostics (pp. 366–374). (1st edn.). Frankfurt: TH-books Verlagsgesellschaft.

Tontonoz P, Hu E, and Spiegelman BM. (1994). Stimulation of adipogenesis in fibroblasts by PPAR gamma 2, a lipid-activated transcription factor. Cell. 79:1147–1156.

Torre, L.C. and Barrit, B.H. (1997). Quantitative evaluation of Rubus fruit anthocyanin pigments. Journal of Food Science, 42, 488–490.

Treutter, D. (1989). Chemical reaction detection of catechins and proanthocyanidins with 4-dimethylaminocinnamaldehyde. Journal of Chromatography 467, 185–193.

Turner, N.D., Diaz, A., Taddeo, S.S., Vanamala, J., McDonough, C.M., Dykes, L., Murphy, M.E., Carroll, R.J. and Rooney, L.W. (2006). Bran from black or brown sorghum suppresses colon carcinogenesis. FASEB Journal 20, A599.

Ummadi, P., Chenoweth, W.L., Uebersax, M.A. (1995). The influence of extrusion processing on iron dialyzability, phytates, and tannins in legumes. Journal of Food Processing and Preservation 19, 119–131.

United States Department of Agriculture – Foreign Agriculture Division, (2003) data. Available from: <http://www.fas.usda.gov/psd>.

United States Grains Council (2001). Sorghum production and usage data. Available from: <http://www.grains.org/grains/sorghum. html>.

USDA National Nutrient Database for Standard Reference, Release 24 (2011). Item # 20648.

Vaaler, S. and Honssen, K.F. (1983). Improvement in long term diabetic control after high fiber (bran and sugar) diets. Diabetologia. 25: 200.

Van Rensburg, S.J. (1981). Epidemiological and dietary evidence for a specific nutritional predisposition to esophageal cancer. Journal of the National Cancer Institute 67, 243–251.

Vidal-Valverde, C., Frias, J., Estrella, M.I., Gorospe, M.J., Ruiz, R. and Bacon, J. (1994). Effects of processing on some antinutritional factors of lentils. Journal of Agricultural and Food Chemistry 42, 2291–2295.

Villaumet, L., Gariot, B., Desalme, P.A. and Debray, G.B. (1984). Long term evolution of the effect of bran ingestion on meat induced glucose and insulin response in healthy man. Am.J. Clin. Nutr. 40:1023-26.

Vohra, P., Kratzer, F.H. and Joslyn, M.A. (1966). The growth depressing and toxic effects of tannins to chicks. Poultry Science 45, 135–142.

Wang, H., Cao, G. and Prior, R.L. (1997). Oxygen radical absorbing capacity of anthocyanins. Journal of Agricultural and Food Chemistry, 45, 304–309.

Wang, J., Kalt, W. and Sporns, P. (2000). Comparison between HPLC and MALDI-TOF MS analysis of anthocyanins in highbush blueberries. Journal of Agricultural and Food Chemistry, 48, 3330–3335.

Waniska, R.D., Hugo, L.F., Rooney, L.W. (1992). Practical methods to determine the presence of tannins in sorghum. Journal of Applied Poultry Research 1, 122–128.

Waniska, R.D., Poe, J.H. and Bandyopadhyay, R. (1989). Effects of growth conditions on grain molding and phenols in sorghum caryopsis. Journal of Cereal Science 10, 217–225.

Waniska, R.D. and Rooney, L.W. (2000). Structure and chemistry of the sorghum caryopsis. In: Smith, C.W., Frederiksen, R.A. (Eds.), Sorghum: Origin, History, Technology, and Production. Wiley, New York, pp. 649–688.

Watanabe, M. (1999). Antioxidative phenolic compounds from Japanese barnyard millet (*Echinochloa utilis*) grains. Journal of Agricultural and Food Chemistry 47, 4500–4505.

Watterson, J.J. and Butler, L. (1983). Occurrence of an unusual leucoanthocyanidin and absence of proanthocyanidin in sorghum leaves. Journal of Agricultural and Food Chemistry 31, 41–45.

Weihrauch, J.L. and Gardner, J.M. (1978). Sterol content of foods of plant origin. Journal of the American Dietetic Association 73, 39–46.

Weller, C.L. and Hwang, K.T. (2003). Recoverable lipids from grain sorghum dry distiller's grain. In: Proceedings of the 23rd Biennial Sorghum Industry Conference, Albuquerque, NM.

Wharton, P.S. and Nicholson, R.L., 2000. Temporal synthesis and radiolabelling of the sorghum 3-deoxyanthocyanidin phytoalexins and the anthocyanin, cyanidin3-dimalonyl glucoside. New Phytologist 145, 457–469.

Willson, T.M., Brown, P.J., Sternbach, D.D. and Henke, B.R. (2000). The PPARs: from orphan receptors to drug discovery. J Med Chem. 43:527–550.

Wrolstad, R.E. (1976). Colour and pigment analyses in fruit products. Station Bulletin 624. Agriculture Experiment Station, Oregon State University.

Wu, X., Beecher, G.R., Holden, J.M., Haytowitz, D.B., Gebhardt, S.E. and Prior, R.L. (2004). Lipophilic and hydrophilic antioxidant capacities of common foods in the United States. Journal of Agricultural and Food Chemistry 52, 4026–4037.

Wu, X., Cao, G. and Prior, R.L. (2002). Absorption and metabolism of anthocyanins in elderly women after consumption of elderberry or blueberry. Journal of Nutrition 132, 1865–1871.

Wu, X. and Prior, R.L. (2005). Identification and characterization of anthocyanins by high-performance liquid chromatography-electrospray ionization-tandem mass spectrometry in common foods in the United States: vegetables, nuts, and grains. Journal of Agricultural and Food Chemistry 53, 3101–3113.

Wyatt, H.R. (2003). The prevalence of obesity. Primary Care 30, 267.

Xu, Z.M. and Godgers, J.S. (2000). Comparison of supercritical fluid and solvent extraction methods in extracting gamma-oryzanol from rice bran. Journal of the American Oil Chemists Society 77, 547–551.

Yang, C.S., Landau, J.M., Huang, M.T. and Newmark, H.L. (2001). Inhibition of carcino-genesis by dietary polyphenolic compounds. Annual Review of Nutrition 21, 381–406.

Yang, T.T.C. and Koo, M.W.L. (2000). Inhibitory effects of Chinese green tea on endothelial cell-induced LDL oxidation. Atherosclerosis 148, 67–73.

Yang, Y. and Chien, M. (2000). Characterization of grape procyanidins using high-performance liquid chromatography/mass spectrometry and matrix-assisted laser desorption/ionization time-of-flight mass spectrometry. Journal of Agricultural and Food Chemistry 48, 3990–3996.

Yasumatsu, K., Nakayama, T.O.M. and Chichester, C.O. (1965). Flavonoids of sorghum. Journal of Food Science 30, 663–667.

Yildiz, A., Karakaplan, M. and Aydin, F. (1998). Studies on *Pleurotus ostreatus* (Jacq. ex Fr.) Kum. var. *salignus* (Pers. ex Fr.) Konr. Et Maubl.: Cultivation, proximate composition, organic and mineral composition of carpophores. Food Chemistry, 61, 127–130.

Yu, C.L. and Swaminathan, B. (1987). Mutagenicity of Proanthocyanidins. Food and Chemical Toxicology 25, 135–139.

Yu, L., Haley, S., Perret, J., Harris, M., Wilson, J. and Qian, M. (2002). Free radical scavenging properties of wheat extracts. Journal of Agricultural and Food Chemistry, 50, 1619–1624.

Zelaya, N., Yeggy, H., Suhendro, E.L., Quintero, X. and Rooney, L.W. (1999). The effect of grain colour and pH on sorghum tortilla chips. AACC Annual Meeting Abstracts. URL: /http://www.aaccnet.org/meetings/99mtg/abstracts/acabc74.htmS

Zielinski, H., and Kozlowska, H. (2000). Antioxidant activity and total phenolics in selected cereal phenolics and their different morphological fractions. Journal of Agricultural and Food Chemistry, 48, 2008–2016

Chapter – 3

Value Added Sorghum Products

PRIMARY PROCESSING

This is the processing of bulk farm produce to clean grain through grading and cleaning by a cleaner-cum-grading machine. This machine grades grain based on the size of the grain and separates it into three sizes, bigger size, medium size, and small size and it also separates unwanted particles. The grain thus separated is collected directly into gunny bags and can be sent to the market. This helps the farmer in getting the correct price for the grain as per its quality. The bulk operation of these processes can be done mechanically.

Destoning (the removal of stones from bulk farm produce) is also another primary processing method of the grain and is carried out by a destoner machine. Destoning is the removal of waste, stones, and glumes from the grain, which is very much necessary as it improves the storage of the grain and consumer acceptability for usage. This is also a kind of value addition to the grain. The percent recovery of the grain from the bulk lots either from the farm or market usually ranges from 89.33-96.60%.

Dehulling or pearling of grain sorghum: It is essential to polish/pearl the grain prior to milling into rawa or flour in a pearling machine. In this process the outer surface of the grain is polished. It may not only remove the bitterness associated with the pericarp, but also reduces insoluble fiber and dirt and microorganisms present on the naked grain, and also improve the appearance of the grain. The mold and fungi that attack the grain grown during the Kharif season (rainy) can blacken the grain. This blackening usually leads to a low market price. Though the grain productivity is greater compared to the postrainy season, the producer ends up with less income due to this blackening problem. Dehulling or pearling is an abrasion technique used to polish the grain. The pearling machine is a modified version of earlier decorticators. Here the grain after polishing is collected on a tray where the bran that is removed is separated by a shaking operation. The bran is also separated during the process and is collected in a container. This bran can be mixed into cattle feed. The recovery of the grain is 95-98% in pearling process.

SECONDARY PROCESSING

For the preparation of any food product the grain should be in the form of flour or in a form that can be conveniently used. Grain was processed into flour, coarse rawa, fine rawa, flakes, or pops through the respective machines. The quality of the products was evaluated by using these products to prepare various foods that are commonly made from rice and wheat. Sorghum flour can be made using a grinding mill called a pulverizer and the flour is sieved to make it fine. The fine flour is used in roti, cake, and biscuit preparations. Sorghum semolina, locally called rawa or suji, can be made in two forms with two different particle sizes, coarse 1.18 mm and fine 0.6 mm, respectively. Both forms of rawa can commercially be made available for market. Rawa recovery will be 50-55% in the case of coarse rawa and 45-50% in the case of fine rawa and the remaining 50% is flour. Rawa is used for preparing Indian culinary preparations such as idli, upma and kesari, and laddu. Fine rawa is made with a Brabender semolina mill. The rawa made in a Brabender semolina machine is made entirely from the endosperm and the pericarp is removed as bran separately. Sorghum flakes (beaten sorghum): Sorghum grain can be processed into flakes or beaten sorghum using the flaking machine. The flaking machine comprises of two parts. One is a roaster which is used for roasting the grain. The other part is a roller flaker used for making the flakes. The flaking machine unit is ideally purposed for rice and the same machine was used for sorghum grain by optimizing the temperature and moisture. The appearance, quality, and taste of these products was found to be acceptable and on a par with rice, wheat, and maize.

SORGHUM SEMOLINA

Improvement in primary process itself has helped in getting fine sorghum flour. Use of appropriate milling procedures and sieves help in getting sorghum semolina. Fine quality sorghum flour can be substituted in place of Bengal gram flour and wheat flour or wheat semolina. The process developed for sorghum flour and semolina is shown in **Figure 1**. In a similar manner, pearl millet, ragi and maize flour and semolina can be obtained.

The grains of ten sorghum varieties *viz.,* Phule Revati, Phule Vasudha, Phule Chitra, Phule Yashoda, Phule Maulee, Phule Anuradha, CSV-22, CSV-18, Selection-3, Maldandi and five hybrids *viz.*, CSH-15-R, SPH-1620, SPH-1647, SPH-1664 and SPH-1665 were obtained from the Senior Sorghum Breeder, All India Co-ordinated Sorghum Improvement Project, Mahatma Phule Krishi Vidyapeeth, Rahuri for the preparation of semolina.

Dehulled sorghum
↓
Coarse milling
↓
Cool
↓
Sieve (40 mesh)
↓
Flour and fine Semolina | Semolina (coarse, yield 28%)

Flour and fine Semolina
↓
Sieve (60 mesh)
↓
Flour (yield 40%) | Semolina (fine) (yield 42%)

Figure 1: Processing of dehulled sorghum in to semolina.

Preparation of semolina from sorghum grains

Semolina (*rava*) from sorghum grains was prepared as per the following procedure (**Figure 2**).

Sorghum grain
↓
Cleaning
↓
Reduction by break rolls
↓
Purification
↓ —— Flour
Granular finished product (semolina/*rava*)
↓
Packaging
↓
Storage

Figure 2: Flow chart for preparation of sorghum semolina (*rava*).

Chemical constituents of sorghum grains and semolina: Chemical analysis of sorghum grains and semolina were done for protein, total sugar, crude fiber, starch, amino acids and minerals using standard methods A.O.A.C., (1990) and NIR Spectrometer, Spectra Analyzer serial No : 05; 281, ZEUTEC Opto Elektronik GmbH, Keiler str. 211, 24768 Rendsburg, Germany.

Preparation of sweet (shira) from sorghum semolina

Recipe for preparation of sorghum sweet (*shira)*: Sorghum semolina 50g, sugar 50g, cashewnut 10g, almond 10g, cardamom 2g, vegetable ghee 25g and water 100ml (as per requirement) (**Figure 3**).

Sorghum *semolina (rava)*

↓

Rosting till it becomes brown

↓

Place deep fry pan on gas

↓

Add ghee and heat to warm

↓

Add sufficient water in it and boil it

↓

Add semolina, sugar, salt etc.

↓

After cooking spread almond, cashewnut

↓

Serve it while hot.

Figure 3: Flow chart for preparation of sorghum sweet (*shira).*

Preparation of *upma* from sorghum semolina

Sorghum semolina 50g, gram flour ½ spoon, blackgram flour ½ spoon, peanuts 20g, tomato 10g, onion 15g, green chilli 2g, mustard 0.2g, curry leaves 2-3 leaves, oil 10g, salt 1g and water 100ml (as per requirement) (**Figure 4**).

Sorghum semolina
↓
Rosting till it becomes brown
↓
Cut onion, chilli, coriander leaves, curry leaves
↓
Place deep fry pan on gas
↓
Add oil and heat to warm
↓
Add mustard, onion, chilli, curry leaves, Gram dhal powder, black gram dhal powder
↓
Fry thoroughly
↓
Add sufficient water in it and boil it
↓
Add semolina, salt (cook 15-20 min.)
↓
After cooking spread coriander leaves on it
↓
Serve it while hot

Figure 4: Flow chart for preparation of sorghum *Upama*

Preparation of *idali* from sorghum semolina

Sorghum semolina 50g, blackgram (dhal) 25g, oil ½ spoon, salt 2g and water 100ml (as per requirement) (**Figure 5**).

Organoleptic evaluation of semolina products

Organoleptic evaluation of sweet (*shira), Upma* and *Idali* for colour and appearance, flavour, texture, taste and overall acceptability was carried out using standard method of Amerine *et al.* (1965). For this 10 semitrained judges were used and 1 to 9 point hedonic scale was used for rating the quality of the sorghum product.

Recovery of semolina (rava) from sorghum grains

Semolina yield ranged from 46.51 to 54.29 per cent. The hybrid genotypes gave higher yield of semolina then the varieties used in the experiment. Hybrid CSH 15R gave significantly highest yield of semolina (54.29%), while among the varieties Selection-3 gave highest semolina yield (50.61%) and at par with M 35-1 (49.57%) followed by Phule Yashoda (49.40%), Phule Revati (49.24%) and Phule Vasudha (49.16%) than other varieties and hybrids (Table 2).

Soak black gram dhal for 8-10 hr

↓

Then drain out the water

↓

Grind into fine paste

↓

soak samolina for 1-2 hr

↓

Add grinded black gram dahl, salt in soaked semolina

↓

Mix thoroughly (Use mixer)

↓

Keep one night for fermentation

↓

Apply teaspoon oil to the *Idali* mould and pour *Idali* batter in it

↓

Steam it in *Idali* cooker for 10-15 min

↓

Take out and remove *Idali* from the moulds

↓

Serve hot with chuteny

Figure 5: Flow chart for preparation of sorghum *Idali*

Chemical constituents of sorghum grain and Semolina

The crude protein content in grain and semolina ranged from 7.81 to 10.45% and 5.43 to 8.35% respectively (**Tables 1 and 2**). Maldandi variety gave significantly higher level of protein (10.45%) in the grain and at par with CSV-22 (10.42%) followed by Selection-3 (10.39%) and Phule Vasudha (10.16%). In semolina, Phule Vasudha gave significantly superior protein (8.35%) followed by CSV-22 (8.06%), Phule Revati (7.61%) and Selection-3 (7.43%) than other. FAO (1995) and Beta *et al.* (1995) was observed content of the protein in whole sorghum grain in the range of 7 to 15 per cent. Robertson and Perez-Maldonado (2006) reported that crude protein in experiment sorghum ranged from 9.14 to 13 per cent. Chavan *et al.* (2009) observed protein content in sorghum ranged from 9.6 to 14 per cent. Similar results observed by Viraktamath *et al.* (1972), Eggum *et al.* (1983) and Ratnavathi *et al.* (2000).

Starch: The starch content in grain and semolina ranged from 61.07-69.01 per cent and 59.93-66.43 per cent, respectively. In the grain Phule Vasudha gave significantly higher level of starch content (69.01%) and at par with Maldandi (68.93%) followed by Phule Yashoda (67.66%), CSV-18 (66.92%) and Phule Revati (65.57%). The statistical analysis showed that the starch content in varieties and hybrids had significant difference. Maldandi gave significantly higher level of starch in semolina (66.43%) followed by CSV-18 (65.05%), Phule Yashoda (64.76%), Phule Vasudha (64.30%) and SPH-1664 (64.20%). The results obtained in the present investigation are in agreement to the literature (Miller and Burns, 1970; Eggum *et al.,* 1983; Ratnavathi *et al.,* 2010; Chavan *et al.,* 2009).

Total sugars: In grain, Phule Yashoda gave significantly higher total sugar (2.32%) followed by Selection-3 (2.12%), CSV-22(1.95%), Phule Revati (1.93%), Phule Maulee (1.93%) and Phule Anuradha (1.93%). In *rava*, Phule Yashoda gave statistically superior higher level of total sugar (2.12%) followed by selection-3 (1.91%), Maldandi (1.75%), Phule Anuradha (1.68%) and CSH-15R (1.61%) (Subramanian and Jambunathan, 1984; Deshpande *et al.,* 2003; Chavan *et al.,* 2009; Ibrahim *et al.,* 2010).

Crude fiber: The crude fiber content in grain and semolina ranged from 2.56-3.41% and 2.44-3.26% respectively. In grain, Phule Maulee gave statistically superior level of crude fiber (3.41%) followed by Phule chitra (3.21%), CSV-22 (3.20%), selection-3 (3.18%) and Phule Anuradha (3.16%). The Statistical analysis showed that there is a significant different in the crude fiber content among the varieties and hybrids. In *rava*, Phule Maulee gave statistically superior level of crude fiber (3.26%) followed Phule Chitra (3.15%), Phule Anuradha (3.13%), CSV-22 (3.12%) and selection-3 (3.12%) (Ratnavathi *et al.,* 2000; Vannalli *et al.,* 2008; Chavan *et al.,* 2009). The crude fiber content in the diet play important role for digestion and bowl movement. It also helps in avoiding constipation problems as well as some other stomach diseases. The results obtained in present investigation are parallel to the literature.

Amino acids content in sorghum grain and semolina: The amino acids content in sorgum grain and semolina are similar (**Tables 3 and 4**). The non-essential amino acids *viz.,* Proline, Alanine, Tyrosine, Glutaminc acid, Glycine, Serine, Aspartic acids threonine, glutamine, asperagine etc. are also present in the sorghum grain. There was a significant difference between the amino acid contents of the varieties and hybrids. The results obtained in the present investigation are in agreement with the literature (Mosse *et al.,* 1988; Robertson *et al.,* 2006; Chavan and Patil, 2010).

Minerals content in sorghum grain and semolina: Calcium content in the sorghum grain ranged from 11.56-27.81 mg/100 g. Selection-3 gave higher level of calcium content (27.81 mg/100 g) followed by Phule chitra (21.54 mg/100 g) and SPH-1665 (20.5 mg/100 g) (**Table 5**). Sorghum grains as well as their semolina are the good source of calcium, iron, phousperous, potassium and other minor elements also. FAO (1995), Chavan and Patil (2010), Winchester and Makokha (2011) reported similar results. The mineral contents in sorghum grain and semolina are slightly different from each other (**Table 6**). This might be due to the processing of sorghum grains into semolina while preparing these products there is a production of flour that also contain mineral elements. Therefore the concentration of these minerals might change slightly. The statistical analysis showed significant difference in the mineral content of grain and semolina within the varieties as well as hybrids.

Organoleptic evaluation of shira, upama and idali prepared from sorghum semolina: Overall acceptability for sweet *(shira)* Phule Vasudha and Phule Yashoda (8.2) gave highest score followed by Phule Chitra (7.6), Maldandi (7.6) and SPH-1620 (7.6). Overall acceptability of *upama* ranged from 6.8 to 8.2. Phule Vasudha had highest (8.2) overall acceptability followed by Phule Yashoda (8.0), CSV-22 (8.0), selection-3 (8.0) and Maldandi gave lowest (6.8) overall acceptability among the all varieties and hybrids. Overall acceptability of *idali* among different varieties and hybrids, score ranged from 7.0 to 8.2. Phule Vasudha and Phule Yashoda (8.2) gave highest overall acceptability followed by CSV-22 (7.8), Phule Revati (7.6), Selection-3 (7.6) and SPH-1664 (7.6) (**Table7**).

While considering the yield of semolina from sorghum grains as well as their nutritional composition and organoleptic properties of the niche products such as sweet (*shira*), *upma* and *idali* prepared from them, the varieties Phule Vasudha and Phule Yashoda are the best one as compared to the other varieties and hybrids. For above all niche products preparation from the sorghum semolina, it was observed that the varieties are better than the hybrids.

Table 1. Nutritional composition of sorghum grains*.

Genotype	Crude protein, %	Starch, %	Total sugars, %	Crude fiber, %
Variety				
Phule Revati (RSV-1006)	9.47	65.57	1.93	2.78
Phule Vasudha (RSV-423)	10.16	69.01	1.63	2.84
Phule Chitra (SPV-1546)	9.74	61.79	1.82	3.21
Phule Yashoda (SPV-1359)	9.49	67.66	2.32	2.82
Phule Maulee (RSLG-262)	9.83	61.27	1.93	3.41
Phule Anuradha (RSV-458)	9.13	62.62	1.93	3.16
CSV-22	10.42	60.38	1.95	3.20
CSV-18	9.45	66.92	1.85	2.76
Selection-3	10.39	61.07	2.12	3.18
Maldandi (M 35-1)	10.45	68.93	1.83	2.92
Hybrid				
CSH-15R	8.75	63.56	1.74	2.83

Genotype	Crude protein,%	Starch, %	Total sugars,%	Crude fiber,%
SPH-1620	8.30	63.71	1.75	2.72
SPH-1647	8.16	63.32	1.45	2.72
SPH-1664	7.81	64.65	1.83	2.56
SPH-1665	8.16	63.55	1.41	2.59
Range	7.81-10.45	61.07-69.01	1.41-2.32	2.56-3.41
Mean	9.31	64.26	1.78	2.91
SE ±	0.014	0.137	0.010	0.021
CD at 5%	0.043	0.398	0.031	0.063
CV%	0.480	0.643	1.823	2.245

*All results are mean values of three determinations.

Table 2. Sorghum semolina recovery and their nutrient contents.

Genotype	Semolina recovery, %	Flour, %	Nutrient content in semolina			
Variety			Crude protein, %	Starch, %	Total sugars, %	Crude fiber, %
Phule Revati (RSV-1006)	49.24	50.76	7.61	64.16	1.05	2.59
Phule Vasudha (RSV-423)	49.16	50.84	8.35	64.30	1.16	2.76
Phule Chitra (SPV-1546)	48.90	51.10	6.73	61.25	1.37	3.15
Phule Yashoda (SPV-1359)	49.40	50.60	6.37	64.76	2.12	2.75
Phule Maulee (RSLG-262)	48.91	51.09	6.64	60.75	1.43	3.26
Phule Anuradha (RSV-458)	48.73	51.27	7.24	61.83	1.68	3.13
CSV-22	46.51	53.49	8.06	59.93	1.46	3.12
CSV-18	48.44	51.56	7.14	65.05	1.33	2.63
Selection-3	50.61	49.39	7.43	60.51	1.91	3.12
Maldandi (M 35-1)	49.57	50.43	7.41	66.43	1.75	2.87
Hybrid						
CSH-15R	54.29	45.71	6.61	63.09	1.61	2.78
SPH-1620	51.32	48.68	6.64	63.04	1.60	2.65
SPH-1647	50.96	49.04	7.26	63.21	1.24	2.63
SPH-1664	52.69	47.31	5.43	64.20	1.42	2.44
SPH-1665	52.38	47.62	6.08	63.20	1.19	2.48

Genotype	Semolina recovery, %	Flour, %	Nutrient content in semolina			
Variety			Crude protein, %	Starch, %	Total sugars, %	Crude fiber, %
Range	46.51-54.29	45.71-53.49	5.43-8.35	59.93-66.43	1.05-2.12	2.44-3.26
Mean	50.08	49.92	7.00	63.04	1.45	2.82
SE ±	0.479	0.475	0.011	0.009	0.007	0.008
CD at 5%	1.385	1.373	0.033	0.028	0.022	0.024
CV%	2.877	2.858	0.500	0.046	1.631	0.893

*All results are mean values of three determinations.

Table 3. Amino acid composition of sorghum grains (g/16 g N)

Genotype	Cys	Glu	Gly	His	Ile	Leu	Lys	Met	Phe
Variety									
Phule Revati (RSV-1006)	0.92	20.77	3.04	2.08	3.80	12.30	2.56	1.38	4.32
Phule Vasudha (RSV-423)	0.94	20.76	2.94	2.11	3.84	12.33	2.52	1.33	4.37
Phule Chitra (SPV-1546)	0.92	20.85	2.93	2.05	3.91	13.22	2.65	1.35	4.44
Phule Yashoda (SPV-1359)	0.95	20.83	3.22	2.06	3.95	12.41	2.50	1.37	4.11
Phule Maulee (RSLG-262)	0.87	20.60	3.06	2.06	3.85	12.77	2.65	1.34	4.45
Phule Anuradha (RSV-458)	0.94	20.45	2.99	2.05	3.92	13.04	2.58	1.36	4.37
CSV-22	0.85	21.03	2.84	2.01	4.02	12.86	2.63	1.40	4.41
CSV-18	0.95	20.55	2.84	2.07	3.82	12.51	2.56	1.38	4.48
Selection-3	0.80	20.85	2.62	2.07	3.94	12.37	2.60	1.39	4.64
Maldandi (M 35-1)	0.87	20.56	2.89	2.09	3.90	13.16	2.52	1.44	4.26
Hybrid									
CSH-15R	0.90	21.05	2.91	2.12	3.90	13.19	2.40	1.38	4.45
SPH-1620	0.96	20.86	3.08	2.08	3.79	12.52	2.49	1.30	4.55
SPH-1647	0.95	20.77	2.84	2.12	3.81	12.98	2.58	1.39	4.35
SPH-1664	0.91	21.17	3.14	2.07	4.13	12.43	2.38	1.35	4.44
SPH-1665	0.95	21.08	3.37	1.98	4.08	12.26	2.36	1.41	4.65
Range	0.80-0.96	20.45-21.17	2.62-3.37	1.98-2.12	3.79-4.13	12.26-13.22	2.36-2.65	1.30-1.44	4.11-4.65
Mean	0.92	20.81	2.98	2.07	3.91	12.69	2.55	1.37	4.42
SE ±	0.007	0.013	0.009	0.006	0.007	0.008	0.011	0.007	0.009
CD at 5% (n=3)	0.021	0.038	0.026	0.017	0.020	0.024	0.033	0.020	0.026
CV%	2.446	0.192	0.933	0.875	0.555	0.201	1.383	1.587	0.616

*All results are mean values of three determinations.

Table. 3 contd.....

Genotype	Pro	Ser	Thr	Trp	Tyr	Ala	Arg	Asp	Val
Variety									
Phule Revati (RSV-1006)	5.93	3.55	2.97	1.17	3.14	8.54	3.65	7.07	3.64
Phule Vasudha (RSV-423)	6.13	3.46	2.95	1.18	3.15	8.67	3.95	6.73	3.77
Phule Chitra (SPV-1546)	5.86	3.49	2.93	1.08	3.10	8.12	3.94	7.29	3.72
Phule Yashoda (SPV-1359)	6.06	3.61	2.94	1.04	3.23	8.48	3.92	7.62	3.74
Phule Maulee (RSLG-262)	5.86	3.56	2.94	1.11	3.04	8.36	3.87	6.55	4.35
Phule Anuradha (RSV-458)	5.92	3.58	2.97	1.14	3.12	8.46	3.15	7.00	3.44
CSV-22	5.93	3.42	2.94	1.17	3.06	8.07	3.76	7.56	4.14
CSV-18	6.22	3.40	2.96	1.16	3.13	8.27	3.69	7.28	3.83
Selection-3	5.83	3.15	2.96	1.20	2.84	8.49	4.57	6.63	4.54
Maldandi (M 35-1)	5.85	3.47	3.06	1.04	3.17	8.27	3.38	7.07	3.86
Hybrid									
CSH-15R	6.36	3.60	2.97	1.07	3.27	8.49	3.37	6.95	3.47
SPH-1620	6.04	3.56	2.93	1.10	3.26	8.85	2.62	6.98	3.73
SPH-1647	6.25	3.37	2.96	1.08	3.25	8.36	3.48	6.95	3.14
SPH-1664	6.24	3.76	2.93	1.13	3.19	8.76	3.78	6.88	3.38
SPH-1665	6.28	3.84	2.97	1.06	3.13	8.55	3.91	6.93	3.22
Range	5.83-6.36	3.15-3.84	2.93-3.06	1.04-1.20	2.84-3.27	8.07-8.85	2.62-4.57	6.55-7.62	3.14-4.54
Mean	6.05	3.52	2.96	1.11	3.14	8.45	3.67	7.03	3.73
SE ±	0.008	0.006	0.008	0.006	0.009	0.006	0.005	0.007	0.007
CD at 5%	0.023	0.019	0.024	0.018	0.027	0.019	0.015	0.020	0.022
CV%	0.410	0.581	0.873	1.720	0.895	0.240	0.433	0.300	0.621

*All results are mean values of three determinations.

Table 4. Amino acid composition of semolina prepared from different genotypes of sorghum (g/16 g N)

Genotype	Cys	Glu	Gly	His	Ile	Leu	Lys	Met	Phe
Variety									
Phule Revati (RSV-1006)	0.75	23.53	2.94	2.53	3.74	14.55	2.07	1.49	4.47
Phule Vasudha (RSV-423)	0.75	23.72	3.15	2.48	3.75	14.57	2.12	1.56	4.46

Genotype	Cys	Glu	Gly	His	Ile	Leu	Lys	Met	Phe
Phule Chitra (SPV-1546)	0.74	24.06	3.05	2.55	3.77	14.94	2.07	1.47	4.55
Phule Yashoda (SPV-1359)	0.88	23.83	3.14	2.56	3.63	14.36	2.13	1.52	4.45
Phule Maulee (RSLG-262)	0.78	23.75	2.95	2.47	3.63	14.73	2.16	1.47	4.47
Phule Anuradha (RSV-458)	0.73	23.60	3.07	2.46	3.94	14.93	2.07	1.55	4.54
CSV-22	0.74	23.84	2.94	2.49	3.85	14.74	2.14	1.49	4.58
CSV-18	0.72	23.57	2.86	2.37	3.74	14.39	2.17	1.47	4.37
Selection-3	0.76	23.24	3.08	2.49	3.83	14.28	2.17	1.57	4.47
Maldandi (M 35-1)	0.74	23.53	3.05	2.55	3.83	14.51	2.08	1.48	4.37
Hybrid									
CSH-15R	0.85	23.95	2.82	2.55	3.63	14.72	2.15	1.46	4.62
SPH-1620	0.85	24.05	2.86	2.56	3.74	15.06	2.06	1.45	4.53
SPH-1647	0.86	23.86	2.94	2.55	3.65	14.34	2.07	1.49	4.47
SPH-1664	0.85	23.65	2.88	2.50	3.73	13.72	2.06	1.45	4.47
SPH-1665	0.76	23.92	2.75	2.52	3.74	14.75	2.11	1.47	4.52
Range	0.72-0.88	23.24-24.06	2.75-3.15	2.37-2.56	3.63-3.94	13.72-15.06	2.06-2.17	1.45-1.57	4.37-4.62
Mean	0.78	23.74	2.96	2.51	3.75	14.57	2.11	1.49	4.49
SE ±	0.006	0.020	0.007	0.008	0.007	0.008	0.005	0.00	0.007
CD at 5%	0.017	0.059	0.022	0.025	0.021	0.024	0.017	0.019	0.021
CV%	2.309	0.261	0.770	1.073	0.592	0.171	0.847	0.365	0.499

*All results are mean values of three determinations.

Table 4 contd.....

Genotype	Pro	Ser	Thr	Trp	Tyr	Ala	Arg	Asp	Val
Variety									
Phule Revati (RSV-1006)	4.92	3.56	3.14	0.75	4.13	7.13	6.05	8.74	5.46
Phule Vasudha (RSV-423)	4.98	3.63	3.14	0.76	4.07	7.05	6.62	8.83	5.63
Phule Chitra (SPV-1546)	5.27	3.60	3.16	0.77	4.06	7.32	6.92	8.36	5.56
Phule Yashoda (SPV-1359)	5.07	3.65	3.13	0.74	4.06	7.15	6.76	8.84	5.28
Phule Maulee (RSLG-262)	5.37	3.45	3.18	0.85	4.06	7.36	6.83	8.24	5.47

Genotype	Pro	Ser	Thr	Trp	Tyr	Ala	Arg	Asp	Val
Phule Anuradha (RSV-458)	4.92	3.74	3.17	0.75	4.09	7.29	6.33	8.52	5.06
CSV-22	4.84	3.64	3.11	0.73	3.92	7.06	6.63	8.82	5.84
CSV-18	5.07	3.47	3.15	0.83	3.90	7.10	5.83	8.64	5.07
Selection-3	5.10	3.74	3.17	0.80	4.04	7.27	5.93	8.53	5.77
Maldandi (M 35-1)	5.14	3.63	3.16	0.70	4.11	7.44	6.26	8.17	5.35
Hybrid									
CSH-15R	5.09	3.47	3.09	0.74	4.05	7.13	6.72	8.37	4.95
SPH-1620	5.35	3.50	3.16	0.73	4.75	7.47	6.77	7.92	5.76
SPH-1647	5.28	3.53	3.08	0.82	3.97	7.38	6.47	8.29	5.36
SPH-1664	5.64	3.48	3.07	0.84	4.04	7.52	6.23	7.63	4.96
SPH-1665	3.39	3.46	3.05	0.72	4.05	7.42	6.55	8.27	5.15
Range	3.39-5.64	3.45-3.74	3.05-3.18	0.70-0.85	3.90-4.75	7.06-7.52	5.83-6.92	7.63-8.84	4.95-5.84
Mean	5.03	4.90	3.13	0.77	4.09	7.27	6.46	8.41	5.38
SE ±	0.006	0.009	0.007	0.007	0.006	0.006	0.009	0.031	0.007
CD at 5%	0.019	0.027	0.021	0.021	0.019	0.019	0.027	0.090	0.022
CV%	0.395	0.802	0.707	2.862	0.506	0.279	0.441	1.112	0.431

*All results are mean values of three determinations.

Table 5. Mineral composition of sorghum grains (mg/100 g)

Genotype	Ca	Fe	Mn	Mg	P	K	Cu	Na	Zn
Variety									
Phule Revati (RSV-1006)	15.47	4.36	2.85	211	491	510	0.80	20.13	3.62
Phule Vasudha (RSV-423)	17.92	4.16	2.94	212	511	534	0.86	21.26	3.66
Phule Chitra (SPV-1546)	21.54	3.98	2.86	211	503	510	0.92	19.36	3.76
Phule Yashoda (SPV-1359)	18.38	3.85	3.06	215	513	516	0.84	21.32	3.76
Phule Maulee (RSLG-262)	18.4	4.63	2.77	223	483	520	0.87	20.07	3.62
Phule Anuradha (RSV-458)	11.56	4.26	2.52	225	496	439	0.86	20.07	3.83
CSV-22	19.82	4.09	2.92	213	503	511	0.87	20.10	3.72
CSV-18	17.29	4.74	2.87	224	504	522	0.83	17.33	3.62

Genotype	Ca	Fe	Mn	Mg	P	K	Cu	Na	Zn
Selection-3	27.81	3.47	2.95	212	518	550	0.87	20.44	3.74
Maldandi (M 35-1)	13.85	4.26	2.84	215	515	491	0.88	21.92	3.63
Hybrid									
CSH-15R	16.29	4.46	2.73	217	501	495	0.95	17.84	3.64
SPH-1620	17.63	4.14	2.95	214	500	529	0.96	20.94	3.54
SPH-1647	13.17	4.76	2.86	225	521	524	0.89	21.32	3.56
SPH-1664	18.55	3.86	2.86	218	521	537	0.86	21.27	3.43
SPH-1665	20.5	3.65	2.85	215	518	527	0.86	23.15	3.61
Range	11.56-27.81	3.47-4.76	2.52-3.06	211-225	483-521	439-550	0.80-0.96	17.33-23.15	3.43-3.83
Mean	17.87	4.18	2.85	217	506	494	0.87	20.43	3.64
SE ±	0.021	0.00	0.007	0.018	0.010	0.008	0.005	0.008	0.008
CD at 5%	0.062	0.021	0.021	0.052	0.029	0.024	0.015	0.025	0.024
CV%	0.364	0.240	0.770	0.02	0.005	0.005	1.847	0.131	0.692

*All results are mean values of three determinations.

Table 6. Mineral composition of semolina prepared from different genotypes of sorghum (mg/100 g)

Genotype	Ca	Fe	Mn	Mg	P	K	Cu	Na	Zn
Variety									
Phule Revati (RSV-1006)	24.06	3.80	4.16	193	570	405	0.96	13.45	4.12
Phule Vasudha (RSV-423)	28.96	3.45	4.09	192	581	381	1.05	15.07	4.34
Phule Chitra (SPV-1546)	31.09	3.33	4.26	192	572	383	1.07	12.64	4.15
Phule Yashoda (SPV-1359)	29.12	2.74	4.31	184	586	397	1.05	15.05	4.25
Phule Maulee (RSLG-262)	29.21	3.47	4.15	192	566	370	1.04	11.82	4.05
Phule Anuradha (RSV-458)	25.59	3.64	3.92	184	567	353	1.06	11.92	4.26
CSV-22	29.37	3.45	4.07	197	580	385	1.04	12.35	4.24
CSV-18	22.82	3.92	4.07	191	582	401	0.95	14.07	4.13
Selection-3	24.77	3.93	3.54	151	564	373	1.05	11.55	4.34
Maldandi (M 35-1)	22.85	3.96	4.07	192	594	383	1.08	12.76	4.12

Genotype	Ca	Fe	Mn	Mg	P	K	Cu	Na	Zn
Hybrid									
CSH-15R	28.72	3.27	4.08	185	576	389	1.04	15.61	4.16
SPH-1620	29.95	3.34	4.13	187	578	370	1.03	11.07	4.15
SPH-1647	26.63	3.83	3.92	196	573	399	1.05	14.13	4.06
SPH-1664	29.42	3.28	3.81	199	592	405	0.96	18.06	3.86
SPH-1665	29.15	3.35	4.13	187	574	378	1.05	15.72	4.06
Range	22.82-31.09	2.74-3.96	3.54-4.31	151-199	564-594	370-405	0.95-1.07	11.07-18.06	3.86-4.34
Mean	27.44	3.52	4.04	188	577	385	1.03	13.68	4.15
SE ±	0.009	0.017	0.007	0.010	0.013	0.009	0.007	0.008	0.009
CD at 5%	0.028	0.050	0.022	0.036	0.037	0.028	0.021	0.024	0.027
CV%	0.106	0.957	0.568	0.017	0.006	0.007	2.163	0.186	0.672

*All results are mean values of three determinations.

Table 7. Overall organoleptic evaluation of sweet, *upama* and *idali* prepared from semolina of different genotypes of sorghum

Genotype	Overall acceptability for sweet (*shira*)	Overall acceptability for *upama*	Overall acceptability for *idali*
Variety			
Phule Revati (RSV-1006)	6.6	7.6	7.6
Phule Vasudha (RSV-423)	8.2	8.2	8.2
Phule Chitra (SPV-1546)	7.6	7.4	7.4
Phule Yashoda (SPV-1359)	8.2	8.0	8.2
Phule Maulee (RSLG-262)	7.2	7.4	7.4
Phule Anuradha (RSV-458)	6.6	7.2	7.0
CSV-22	6.8	8.0	7.8
CSV-18	7.4	7.4	7.2
Selection-3	7.4	8.0	7.6
Maldandi (M 35-1)	7.6	6.8	7.0
Hybrid			
CSH-15R	7.2	7.0	7.4
SPH-1620	7.6	7.8	7.4
SPH-1647	7.4	7.0	7.0
SPH-1664	7.2	7.4	7.6
SPH-1665	7.4	7.8	7.4

Genotype	Overall acceptability for sweet (*shira*)	Overall acceptability for *upama*	Overall acceptability for *idali*
Range	6.6-8.2	6.8-8.2	7.0-8.2
Mean	7.36	7.52	7.48
SE ±	0.101	0.128	0.117
CD at 5%	0.288	0.363	0.333
CV%	6.928	8.549	7.871

*All results are mean values of ten determinations. Semitrainde judges and 1 to 9 hedonic scales were used.

Sorghum flakes

Sorghum is a coarse grain and is difficult to pound. Therefore, it was not as popular as rice flakes. But with the advancement in the technology, it has now become possible to produce flakes from sorghum as well. It is a very popular cereal especially in rural areas of western India. Sorghum flakes have many applications. The preferred locations are Maharashtra, Madhya Pradesh and Karnataka. They are used for making deep fat fried products such as fried seasoned mixtures or toasted and seasoned mixtures, energy food, upma, sweet/ savoury pongal, sweet gravy, tamarind bhaat etc.

Ten sorghum varieties *viz.,* Phule Revati, Phule Vasudha, Phule Chitra, Phule Yashoda, Phule Maulee, Phule Anuradha, CSV-22, CSV-18, Selection-3, Maldandi and five hybrids *viz.*, CSH-15-R, SPH-1620, SPH-1647, SPH-1664 and SPH-1665 were obtained from the Senior Sorghum Breeder, All India Co-ordinated Sorghum Improvement Project, Mahatma Phule Krishi Vidyapeeth, Rahuri for preparation of sorghum flakes (**Figure 6**).

Preparation of sorghum flakes and their products

Preparation of sorghum flakes Chiwada: Recipe for the preparation of sorghum flakes chiwada: flakes 50g, oil 10g, turmeric powder 0.5g, chanadhal 5g, peanuts 10g, mustard and cumin 0.2g, chilli powder 2g, curry leaves 2-3 leaves and salt 1g. Chiwada was prepared as per following method (**Figure 7**).

Preparation of sorghum Poha Recipe for sorghum poha: Flakes 50g, peanuts 10g, oil 10g, onion 10g, coriander leaves 2g, salt 1g, green chilli 2g, cumin 1g and mustard 1g. Preparation procedure is as follows (**Figure 8**).

Organoleptic evaluation of Chiwada and Poha: Organoleptic evaluation of *Chiwada* and *Poha* for colour and appearance, flavour, texture, taste and overall acceptability was carried out using standard method of Amerine *et al.* (1965). For this 10 semi-trained judges were used and 1 to 9 point hedonic scale was used for rating the quality of the sorghum product.

Sorghum grains

↓

Cleaning

↓

Sun drying (5-6 hr)

↓

Washing (2-3 times)

↓

Conditioning (12 hr in closed stainless steel drum)

↓

Roasting (340°C for 1 min)

↓

Flaking in flaking machine

↓ → Husk separated

Sieving

↓ → Broken seperated

Flakes

↓

Packaging storage

Figure 6: Flow chart for preparation of sorghum *flakes*

Sorghum flakes

↓

Fry in sufficient quantity of oil

↓

Fry cumin, mustard, Chanadahl, groundnut,
curry leaves, chilli powder, turmeric powder in other pan

↓

Add fried flakes in it and mix well

↓

Serve it while hot.

Figure 7: Flow chart for preparation of sorghum *Chiwada*

Sorghum flakes

↓

Sprinkle small quantity of water on sorghum flakes to moisture it

↓

Soak for 1-2 min.

↓

Fry cumin, mustard, groundnut, curry leaves, green chilli (chopped), turmeric powder in oil

↓

Add moistured flakes in it and mix well

↓

Serve it while hot.

Figure 8: Flow chart for preparation of sorghum *Poha*

Recovery of flakes from sorghum grains: The *flakes* yield was ranged from 50.62 to 64.14 per cent. The Phule Anuradha gave significantly highest yield 64.14 per cent of *flakes* and at par with Phule Maulee (63.43%) and followed by Phule Chitra (62.42%) and M 35-1 (62.32%) (**Table 8**). While preparation of *flakes* from various sorghum genotypes the other various parts are classified as broken and husk.

Chemical constituents of sorghum grain and flakes: The crude protein content in grain and *flakes* ranged from 7.81 to 10.45% and 6.89 to 9.71% respectively. Maldandi variety gave significantly higher level of protein (10.45%) in the grain and at par with CSV-22 (10.42%) followed by Selection-3 (10.39%) and Phule Vasudha (10.16%). FAO (1995) and Beta *et al.* (1995) was observed content of the protein in whole sorghum grain in the range of 7 to 15 per cent (Robertson *et al.,* 2006; Chavan *et al.,* 2010; Viraktamath *et al.,* 1972; Eggum *et al.,* 1983 and Ratnavathi *et al.,* 2000). In *flakes*, Maldandi variety gave statistically superior in protein (9.71%) followed by Selection-3 (9.62%), Phule Vasudha (9.25%), CSV-22 (9.12%) and Phule Revati (8.83) than other hybrids and varieties (**Tables 9 and 10**).

Starch: The starch content in grain and *flakes* ranged from 61.07-69.01 and 59.95-67.87 per cent respectively. In the grain Phule Vasudha gave significantly higher level of starch content (69.01%) and at par with Maldandi (68.93%) followed by Phule Yashoda (67.66%), CSV-18 (66.92%) and Phule Revati (65.57%). The statistical analysis showed that the starch content in varieties and hybrids had significant difference (Miller and Burns, 1970; Eggum *et al.,* 1983; Ratnavathi *et al.,* 2010; Chavan *et al.,* 2009). In *flakes* Maldandi gave statistically superior level of starch (67.87%) followed by Phule Vasudha (67.68%), Phule Yashoda (66.34%), CSV-18 (65.52%) and Phule Revati (65.11%).

Total sugars: Total sugar contain in grain and *flakes* ranged from 1.41-2.32 and 1.06-1.88 per cent, respectively. In grain, Phule Yashoda gave significantly higher total sugar (2.32%) followed by Selection-3 (2.12%), CSV-22(1.95%), Phule Revati (1.93%), Phule Maulee (1.93%) and Phule Anuradha (1.93%) (Subramanian and Jambunathan, 1984; Deshpande *et al.,* 2003; Chavan *et al.,* 2009; Ibrahim *et al.,* 2010). In *flakes* (**Table 10**), selection-3 gave significantly higher level of total sugar (1.88%) and at par with Phule Yashoda (1.87%) followed by CSV-22 (1.74%), Maldandi (1.74%) and Phule Maulee (1.73%).

Crude fiber: The crude fiber content in grain and *flakes* ranged from 2.56-3.41% and 2.14-3.12% respectively (Ratnavathi *et al.,* 2000; Vannalli *et al.,* 2008; Chavan *et al.,* 2009). In *flakes* Phule Maulee gave significantly higher level of crude fiber (3.12%) and at par with CSV-22 (3.10%) followed by Phule Anuradha (3.04%), selection-3 (2.95%) and Phule Chitra (2.93%). The fiber content in the sorghum *flakes* showed lower range than the grains due to the loss during processing.

Amino acids content in sorghum grain and flakes: There was a significant difference between the amino acid contents of the varieties, hybrids and flakes. This difference might be due to the processing effect. The non-essential amino acids *viz.,* proline, alanine, tyrosine, glutaminc acid, glycine, serine, aspartic acids, threonine, glutamine, asperagine etc. are also present in the sorghum grain and flakes (**Tables 11 & 12**). The results obtained in the present investigation are in agreement with the literature (Mosse *et al.,* 1988; Robertson *et al.,* 2006; Chavan and Patil, 2010).

Minerals content in sorghum grain and flakes: Calcium content in the sorghum grain ranged from 11.56-27.81 mg/100 g (**Tables 13 and 14**). Selection-3 gave higher level of calcium content (27.81 mg/100 g) followed by Phule chitra (21.54 mg/100 g) and SPH-1665 (20.5 mg/100 g) (FAO, 1995; Chavan and Patil, 2010; Winchester and Makokha, 2011). Iron content in the sorghum grain ranged from 3.47 to 4.76 mg/100 g. Phule Anuradha gave higher level of Iron content (10.26 mg/100 g) followed by SPH-1647 (9.76 mg/100 g) and CSV-18 (9.74 mg/100 g). Chavan and Patil (2010) observed iron content in the sorghum grain was 4.48 mg/100 g. There is no much research work on the sorghum *flakes.* Therefore the literature is also not available on preparation of sorghum *flakes* and their nutritional composition. So the attempt has been made to prepare sorghum *flakes* and to find out their nutritional value. The mineral contents in sorghum grain and *flakes* are slightly different from each other. This might be due to the processing of sorghum grains into *flakes,* while preparing these products there is a production of flour, brokens and husk that also contain mineral elements. Therefore the concentration of these minerals might change slightly. The statistical analysis showed significant difference in the mineral content of grain, *flakes* and within the varieties as well as hybrids.

Organoleptic evaluation of *chiwada* and *poha* prepared from flakes of sorghum grains

Chiwada: Overall acceptability score of *chiwada* ranged from 6.4-8.0 (**Table 15**). Phule Vasudha and Phule Yasodha gave highest score (8.0) followed by Phule Revati (7.8), Phule Chitra (7.8), CSV-22 (7.8) Selection-3 (7.8) and SPH-1665 (7.8). Overall acceptability of *poha*, score ranged from 6.6-8.2 (Table 8). Selection-3 had highest overall acceptability (8.2) than other varieties and hybrids followed by Phule Vasudha (8.0), Phule Yashoda (8.0), Phule Chitra (7.8), CSV-22 (7.8), CSV-18 (7.8), SPH-1620 (7.8) and SPH-1664 (7.8).

Table 8: Recovery of *flakes* from different genotypes of sorghum.

Genotype	*Flakes* (%)	Broken (%)	Husk (%)
Variety			
Phule Revati (RSV-1006)	53.55	18.32	28.13
Phule Vasudha (RSV-423)	54.60	17.46	27.94
Phule Chitra (SPV-1546)	62.42	10.79	26.79
Phule Yashoda (SPV-1359)	58.78	14.35	26.87
Phule Maulee (RSLG-262)	63.43	13.56	23.01
Phule Anuradha (RSV-458)	64.14	8.45	27.41
CSV-22	51.73	18.41	29.86
CSV-18	50.62	17.52	31.86
Selection-3	51.89	18.35	29.76
Maldandi (M 35-1)	62.32	9.79	27.89
Hybrid			
CSH-15R	61.04	17.49	21.47
SPH-1620	53.37	22.40	24.23
SPH-1647	60.03	14.32	25.65
SPH-1664	61.76	15.25	22.99
SPH-1665	58.55	19.04	22.41
Range	50.62-64.14	8.45-22.40	21.47-31.86
Mean	57.88	15.71	26.41
SE ±	0.436	0.315	0.298
CD at 5% (n=3)	1.261	0.911	0.863
CV%	2.264	6.027	3.395

There is no research work done on the sorghum *flakes* and their products such as *chiwada* and *poha* preparation as well as their organoleptic properties studies. So no literature on this aspect is available. In the present research work, we tried to prepare products and also judge the consumers acceptability.

Therefore, these sorghum value added products should be popularized through various media and make awareness among the consumers is very essential.

Table 9: Nutritional composition of sorghum grains.

Genotype	Crude protein, %	Starch, %	Total sugars, %	Crude fiber, %
Variety				
Phule Revati (RSV-1006)	9.47	65.57	1.93	2.78
Phule Vasudha (RSV-423)	10.16	69.01	1.63	2.84
Phule Chitra (SPV-1546)	9.74	61.79	1.82	3.21
Phule Yashoda (SPV-1359)	9.49	67.66	2.32	2.82
Phule Maulee (RSLG-262)	9.83	61.27	1.93	3.41
Phule Anuradha (RSV-458)	9.13	62.62	1.93	3.16
CSV-22	10.42	60.38	1.95	3.20
CSV-18	9.45	66.92	1.85	2.76
Selection-3	10.39	61.07	2.12	3.18
Maldandi (M 35-1)	10.45	68.93	1.83	2.92
Hybrid				
CSH-15R	8.75	63.56	1.74	2.83
SPH-1620	8.30	63.71	1.75	2.72
SPH-1647	8.16	63.32	1.45	2.72
SPH-1664	7.81	64.65	1.83	2.56
SPH-1665	8.16	63.55	1.41	2.59
Range	7.81-10.45	61.07-69.01	1.41-2.32	2.56-3.41
Mean	9.31	64.26	1.78	2.91
SE ±	0.014	0.137	0.010	0.021
CD at 5% (n=3)	0.043	0.398	0.031	0.063
CV%	0.480	0.643	1.823	2.245

While considering the yield of *flakes* from sorghum grains as well as their nutritional composition and organoleptic properties of the niche products prepared from them, the varieties Phule Vasudha and Phule Yashoda are the best one as compared to the other varieties and hybrids. For above all niche products preparation from the sorghum *flakes,* it was observed that the varieties are better than the hybrids.

Table 10: Nutrient content in sorghum *flakes.*

Genotype	Crude protein (%)	Starch (%)	Total sugars (%)	Crude fiber (%)
Variety				
Phule Revati (RSV-1006)	8.83	65.11	1.06	2.14
Phule Vasudha (RSV-423)	9.25	67.68	1.54	2.34
Phule Chitra (SPV-1546)	8.45	61.26	1.62	2.93
Phule Yashoda (SPV-1359)	8.70	66.34	1.87	2.54
Phule Maulee (RSLG-262)	8.11	60.81	1.73	3.12
Phule Anuradha (RSV-458)	8.19	61.55	1.63	3.04
CSV-22	9.12	59.95	1.74	3.10
CSV-18	8.82	65.52	1.64	2.62
Selection-3	9.62	60.56	1.88	2.95
Maldandi (M 35-1)	9.71	67.87	1.74	2.86
Hybrid				
CSH-15R	8.53	63.35	1.59	2.76
SPH-1620	8.13	63.42	1.59	2.67
SPH-1647	7.92	63.17	1.22	2.63
SPH-1664	6.89	64.62	1.53	2.49
SPH-1665	7.34	63.06	1.20	2.52
Range	6.89-9.71	59.95-67.87	1.06-1.88	2.14-3.12
Mean	8.50	63.62	1.53	2.71
SE ±	0.008	0.009	0.007	0.008
CD at 5% (n=3)	0.023	0.027	0.022	0.024
CV%	0.285	0.045	1.498	0.950

Table 11: Amino acid composition of sorghum grains (g/16 g N)

Genotype	Cys	Glu	Gly	His	Ile	Leu	Lys	Met	Phe
Variety									
Phule Revati (RSV-1006)	0.92	20.77	3.04	2.08	3.80	12.30	2.56	1.38	4.32
Phule Vasudha (RSV-423)	0.94	20.76	2.94	2.11	3.84	12.33	2.52	1.33	4.37
Phule Chitra (SPV-1546)	0.92	20.85	2.93	2.05	3.91	13.22	2.65	1.35	4.44
Phule Yashoda (SPV-1359)	0.95	20.83	3.22	2.06	3.95	12.41	2.50	1.37	4.11
Phule Maulee (RSLG-262)	0.87	20.60	3.06	2.06	3.85	12.77	2.65	1.34	4.45
Phule Anuradha (RSV-458)	0.94	20.45	2.99	2.05	3.92	13.04	2.58	1.36	4.37
CSV-22	0.85	21.03	2.84	2.01	4.02	12.86	2.63	1.40	4.41
CSV-18	0.95	20.55	2.84	2.07	3.82	12.51	2.56	1.38	4.48
Selection-3	0.80	20.85	2.62	2.07	3.94	12.37	2.60	1.39	4.64
Maldandi (M 35-1)	0.87	20.56	2.89	2.09	3.90	13.16	2.52	1.44	4.26
Hybrid									
CSH-15R	0.90	21.05	2.91	2.12	3.90	13.19	2.40	1.38	4.45
SPH-1620	0.96	20.86	3.08	2.08	3.79	12.52	2.49	1.30	4.55
SPH-1647	0.95	20.77	2.84	2.12	3.81	12.98	2.58	1.39	4.35
SPH-1664	0.91	21.17	3.14	2.07	4.13	12.43	2.38	1.35	4.44
SPH-1665	0.95	21.08	3.37	1.98	4.08	12.26	2.36	1.41	4.65
Range	0.80-0.96	20.45-21.17	2.62-3.37	1.98-2.12	3.79-4.13	12.26-13.22	2.36-2.65	1.30-1.44	4.11-4.65
Mean	0.92	20.81	2.98	2.07	3.91	12.69	2.55	1.37	4.42
SE ±	0.007	0.013	0.009	0.006	0.007	0.008	0.011	0.007	0.009
CD at 5% (n=3)	0.021	0.038	0.026	0.017	0.020	0.024	0.033	0.020	0.026
CV%	2.446	0.192	0.933	0.875	0.555	0.201	1.383	1.587	0.616

Table 11: contd.....

Genotype	Pro	Ser	Thr	Trp	Tyr	Ala	Arg	Asp	Val
Variety									
Phule Revati (RSV-1006)	5.93	3.55	2.97	1.17	3.14	8.54	3.65	7.07	3.64
Phule Vasudha (RSV-423)	6.13	3.46	2.95	1.18	3.15	8.67	3.95	6.73	3.77
Phule Chitra (SPV-1546)	5.86	3.49	2.93	1.08	3.10	8.12	3.94	7.29	3.72
Phule Yashoda (SPV-1359)	6.06	3.61	2.94	1.04	3.23	8.48	3.92	7.62	3.74
Phule Maulee (RSLG-262)	5.86	3.56	2.94	1.11	3.04	8.36	3.87	6.55	4.35
Phule Anuradha (RSV-458)	5.92	3.58	2.97	1.14	3.12	8.46	3.15	7.00	3.44
CSV-22	5.93	3.42	2.94	1.17	3.06	8.07	3.76	7.56	4.14
CSV-18	6.22	3.40	2.96	1.16	3.13	8.27	3.69	7.28	3.83
Selection-3	5.83	3.15	2.96	1.20	2.84	8.49	4.57	6.63	4.54
Maldandi (M 35-1)	5.85	3.47	3.06	1.04	3.17	8.27	3.38	7.07	3.86
Hybrid									
CSH-15R	6.36	3.60	2.97	1.07	3.27	8.49	3.37	6.95	3.47
SPH-1620	6.04	3.56	2.93	1.10	3.26	8.85	2.62	6.98	3.73
SPH-1647	6.25	3.37	2.96	1.08	3.25	8.36	3.48	6.95	3.14
SPH-1664	6.24	3.76	2.93	1.13	3.19	8.76	3.78	6.88	3.38
SPH-1665	6.28	3.84	2.97	1.06	3.13	8.55	3.91	6.93	3.22
Range	5.83-6.36	3.15-3.84	2.93-3.06	1.04-1.20	2.84-3.27	8.07-8.85	2.62-4.57	6.55-7.62	3.14-4.54
Mean	6.05	3.52	2.96	1.11	3.14	8.45	3.67	7.03	3.73
SE ±	0.008	0.006	0.008	0.006	0.009	0.006	0.005	0.007	0.007
CD at 5% (n=3)	0.023	0.019	0.024	0.018	0.027	0.019	0.015	0.020	0.022
CV%	0.410	0.581	0.873	1.720	0.895	0.240	0.433	0.300	0.621

Table 12: Amino acid composition of *flakes* prepared from different genotypes of sorghum (g/16 g N)

Genotype	Cys	Glu	Gly	His	Ile	Leu	Lys	Met	Phe
Variety									
Phule Revati (RSV-1006)	1.08	22.63	3.25	2.74	3.86	14.76	1.95	1.54	3.64
Phule Vasudha (RSV-423)	0.97	22.75	3.25	2.74	3.76	14.43	1.93	1.47	3.85
Phule Chitra (SPV-1546)	1.13	22.12	3.61	2.74	3.36	12.54	1.92	1.46	3.25
Phule Yashoda (SPV-1359)	1.07	22.33	3.29	2.73	3.65	13.92	1.93	1.47	3.55
Phule Maulee (RSLG-262)	1.06	21.75	3.48	2.63	3.68	13.07	1.88	1.45	3.24
Phule Anuradha (RSV-458)	1.09	21.54	3.22	2.63	3.85	13.82	1.82	1.45	3.37
CSV-22	1.04	22.51	3.25	2.70	3.73	14.06	2.06	1.55	3.77
CSV-18	1.06	22.73	3.06	2.72	3.83	14.54	1.91	1.54	3.84
Selection-3	1.09	22.77	3.12	2.60	3.96	14.76	2.00	1.55	3.96
Maldandi (M 35-1)	0.97	22.76	3.03	2.76	3.95	15.06	1.96	1.47	3.87
Hybrid									
CSH-15R	1.02	22.05	3.46	2.62	3.65	13.06	1.95	1.48	3.45
SPH-1620	0.77	22.62	2.72	2.64	4.45	15.93	2.07	1.48	4.23
SPH-1647	0.96	22.27	3.15	2.68	3.84	14.57	1.85	1.44	3.64
SPH-1664	0.95	21.97	3.33	2.66	3.85	13.69	1.98	1.44	3.60
SPH-1665	0.96	22.29	3.12	2.68	3.95	14.14	1.96	1.44	3.72
Range	0.77-1.13	21.54-22.77	2.72-3.61	2.60-2.76	3.36-4.45	12.54-15.93	1.82-2.07	1.44-1.55	3.24-4.23
Mean	1.01	22.34	3.22	2.69	3.82	14.16	1.95	1.48	3.66
SE ±	0.005	0.009	0.007	0.005	0.008	0.007	0.005	0.006	0.007
CD at 5% (n=3)	0.015	0.028	0.020	0.015	0.024	0.022	0.015	0.019	0.021
CV%	1.577	0.130	0.659	0.614	0.656	0.163	0.833	1.350	0.594

Table 12: Contd…..

Genotype	Pro	Ser	Thr	Trp	Tyr	Ala	Arg	Asp	Val
Variety									
Phule Revati (RSV-1006)	6.86	3.56	3.15	1.06	4.74	7.91	7.06	4.03	4.57

Genotype	Pro	Ser	Thr	Trp	Tyr	Ala	Arg	Asp	Val
Phule Vasudha (RSV-423)	6.57	3.46	3.15	1.07	4.58	7.83	7.25	4.13	4.54
Phule Chitra (SPV-1546)	7.84	3.77	3.09	1.26	4.81	8.58	7.74	2.56	4.95
Phule Yashoda (SPV-1359)	7.15	3.57	3.06	1.15	4.64	8.35	7.06	3.34	4.64
Phule Maulee (RSLG-262)	7.77	3.64	3.10	1.32	4.62	8.63	6.14	2.40	5.07
Phule Anuradha (RSV-458)	7.54	3.66	2.96	1.25	4.73	8.75	7.15	2.72	4.75
CSV-22	6.83	3.67	3.12	1.17	4.55	7.94	7.35	3.85	4.36
CSV-18	6.77	3.65	3.16	1.07	4.64	7.85	7.15	2.58	4.47
Selection-3	6.45	3.57	3.17	1.06	4.56	7.76	7.06	4.13	4.60
Maldandi (M 35-1)	6.64	3.64	3.03	1.06	4.63	7.86	7.37	3.75	4.34
Hybrid									
CSH-15R	7.42	3.57	3.13	1.05	4.66	8.42	7.15	2.74	5.04
SPH-1620	6.23	3.65	3.17	1.27	4.15	7.73	7.17	4.02	4.61
SPH-1647	6.93	3.43	3.03	1.05	1.65	8.14	6.46	3.62	4.62
SPH-1664	7.15	3.63	3.06	1.14	4.65	8.36	7.15	2.92	4.83
SPH-1665	6.97	3.58	3.04	1.25	4.49	8.22	7.21	3.35	4.63
Range	6.23-7.84	3.43-3.77	2.96-3.17	1.05-1.32	1.65-4.81	7.73-8.75	6.14-7.74	2.40-4.13	4.34-5.07
Mean	7.01	3.6	3.09	1.15	4.40	8.15	7.10	3.34	4.67
SE ±	0.007	0.006	0.006	0.007	0.008	0.006	0.007	0.006	0.007
CD at 5% (n=3)	0.021	0.019	0.020	0.021	0.025	0.019	0.022	0.018	0.022
CV%	0.323	0.549	0.671	1.902	0.592	0.242	0.326	0.584	0.510

Table 13: Mineral composition of sorghum grains (mg/100 g)

Genotype	Ca	Fe	Mn	Mg	P	K	Cu	Na	Zn
Variety									
Phule Revati (RSV-1006)	15.47	4.36	2.85	211	491	510	0.80	20.13	3.62
Phule Vasudha (RSV-423)	17.92	4.16	2.94	212	511	534	0.86	21.26	3.66
Phule Chitra (SPV-1546)	21.54	3.98	2.86	211	503	510	0.92	19.36	3.76
Phule Yashoda (SPV-1359)	18.38	3.85	3.06	215	513	516	0.84	21.32	3.76

Genotype	Ca	Fe	Mn	Mg	P	K	Cu	Na	Zn
Phule Maulee (RSLG-262)	18.4	4.63	2.77	223	483	520	0.87	20.07	3.62
Phule Anuradha (RSV-458)	11.56	4.26	2.52	225	496	439	0.86	20.07	3.83
CSV-22	19.82	4.09	2.92	213	503	511	0.87	20.10	3.72
CSV-18	17.29	4.74	2.87	224	504	522	0.83	17.33	3.62
Selection-3	27.81	3.47	2.95	212	518	550	0.87	20.44	3.74
Maldandi (M 35-1)	13.85	4.26	2.84	215	515	491	0.88	21.92	3.63
Hybrid									
CSH-15R	16.29	4.46	2.73	217	501	495	0.95	17.84	3.64
SPH-1620	17.63	4.14	2.95	214	500	529	0.96	20.94	3.54
SPH-1647	13.17	4.76	2.86	225	521	524	0.89	21.32	3.56
SPH-1664	18.55	3.86	2.86	218	521	537	0.86	21.27	3.43
SPH-1665	20.5	3.65	2.85	215	518	527	0.86	23.15	3.61
Range	11.56-27.81	3.47-4.76	2.52-3.06	211-225	483-521	439-550	0.80-0.96	17.33-23.15	3.43-3.83
Mean	17.87	4.18	2.85	217	506	494	0.87	20.43	3.64
SE ±	0.021	0.00	0.007	0.018	0.010	0.008	0.005	0.008	0.008
CD at 5% (n=3)	0.062	0.021	0.021	0.052	0.029	0.024	0.015	0.025	0.024
CV%	0.364	0.240	0.770	0.02	0.005	0.005	1.847	0.131	0.692

Table 14: Mineral composition of *flakes* prepared from different genotypes of sorghum (mg/100 g).

Genotype	Ca	Fe	Mn	Mg	P	K	Cu	Na	Zn
Variety									
Phule Revati (RSV-1006)	15.13	4.85	4.46	178	574	262	0.96	6.63	3.5
Phule Vasudha (RSV-423)	17.37	5.07	4.38	173	536	256	0.96	5.55	3.64
Phule Chitra (SPV-1546)	17.13	3.39	4.86	145	540	312	0.75	8.97	3.39
Phule Yashoda (SPV-1359)	15.63	4.84	4.45	169	539	264	0.85	6.57	3.57
Phule Maulee (RSLG-262)	17.37	3.83	4.53	148	539	292	0.75	4.58	3.47
Phule Anuradha (RSV-458)	12.56	5.12	3.93	158	510	234	0.75	1.54	3.50
CSV-22	18.24	4.54	4.47	172	545	286	0.94	1.47	3.64

Genotype	Ca	Fe	Mn	Mg	P	K	Cu	Na	Zn
CSV-18	15.73	5.12	4.45	182	546	276	0.97	7.19	3.56
Selection-3	17.07	5.46	4.32	186	535	249	0.96	4.07	3.75
Maldandi (M 35-1)	13.72	5.17	4.33	182	547	266	0.93	5.63	3.57
Hybrid									
CSH-15R	16.92	4.15	4.53	158	529	296	0.73	5.93	3.48
SPH-1620	24.07	5.96	3.84	187	501	261	1.06	2.48	3.84
SPH-1647	14.63	5.57	4.38	177	525	263	0.85	3.16	3.61
SPH-1664	15.72	5.06	4.37	169	515	270	0.83	4.14	3.55
SPH-1665	19.18	4.85	4.33	171	518	272	0.88	2.09	3.60
Range	12.56-24.07	3.39-5.96	3.84-4.86	158-187	501-574	234-312	0.73-1.06	1.47-8.97	3.39-3.84
Mean	16.69	4.86	4.37	170	533	271	0.88	4.67	3.57
SE ±	0.007	0.051	0.008	0.009	0.008	0.016	0.007	0.007	0.008
CD at 5% (n=3)	0.022	0.144	0.024	0.027	0.025	0.046	0.021	0.021	0.023
CV%	0.141	3.167	0.572	0.016	0.004	0.018	2.46	0.487	0.671

Table 15: Overall organoleptic evaluation of *chiwada* and *poha* prepared from *flakes* of different genotypes of sorghum

Genotype	Overall acceptability for *Chiwada*	Overall acceptability for *Poha*
Variety		
Phule Revati (RSV-1006)	7.8	7.6
Phule Vasudha (RSV-423)	8.0	8.0
Phule Chitra (SPV-1546)	7.8	7.8
Phule Yashoda (SPV-1359)	8.0	8.0
Phule Maulee (RSLG-262)	7.4	7.0
Phule Anuradha (RSV-458)	7.6	7.2
CSV-22	7.8	7.8
CSV-18	7.6	7.8
Selection-3	7.8	8.2
Maldandi (M 35-1)	6.4	6.8
Hybrid		
CSH-15R	7.2	7.6
SPH-1620	7.4	7.8
SPH-1647	7.0	6.6

Genotype	Overall acceptability for *Chiwada*	Overall acceptability for *Poha*
SPH-1664	7.6	7.8
SPH-1665	7.8	7.6
Range	6.4-8.0	6.6-8.2
Mean	7.52	7.54
SE ±	0.129	0.146
CD at 5% (n=5)	0.366	0.415
CV%	8.618	9.737

Sorghum protein digestibility

The human feeding trials results underlined the need for a better understanding of sorghum food preparation so that the protein present in the grain may be made fully available in the food prepared from it. Cooked maize and millet showed protein digestibilties of 82 – 87 percent, but that for ordinary sorghum was only 56 percent. However, high-lysine sorghum (P721-Q) gave the better level of 73 percent protein digestibility. This difference in these two types of sorghum genotype may be due to the prolamin proteins were the main source of protein indigestibility in normal sorghums, and they suspected that the problem lay in the formation of protein polymers linked by disulphide bonds. By cooking normal sorghum in the presence of an appropriate reducing agent, protein digestibility increased 25 percent, whereas there was no improvement in barley, rice or wheat when cooked in this way. Reducing agents found suitable included 2-mercaptoethanol, dithiothreitol, sodium bisulphate, and L-cysteine.

Fermentation

Pre-fermentation processing of cereals: Pre-fermentation processing of cereals is largely dependent on the end product desired. In most cases, grains are sun dried prior to fermentation. Treatments such as washing, steeping, milling and sieving are pre-fermentation processing steps applied in the cereals.

Fermented cereals: Plant tissues or animal tissues subjected to the action of microorganisms and or enzymes to give desired biochemical changes and significant modification of food quality are referred to as fermented foods. Fermentation is the oldest known form of food biotechnology; records of barley conversion to beer date back than 5000 years. According to Steinkraus the traditional fermentation of foods serves several functions.

1. Enrichment of the diet through development of a diversity of flavours, aromas and textures in food substrates.
2. Preservation of substantial amounts of food through lactic acid, alcoholic, acetic acid, and alkaline fermentations.

3. Enrichment of food substrates biologically with protein, essential amino acid, essential fatty acids, and vitamins.
4. Detoxification during food fermentation processing.
5. A decrease in cooking times and fuel requirements.
6. Easily digestible and available quickly to the body.

Aside from alcoholic fermentations and the production of yogurt and leavened bread, food fermentations continue to be important primarily in developing countries where the lack of resources limits the use of techniques such as vitamin enrichment of foods, and the use of energy and capital-intensive processes for food preservation. The technology of producing many indigenous fermented foods from cereals remains a household art in these countries. Prospects for applying advanced technologies to indigenous fermented foods and for the production of value added additive products such as colours; flavours, enzymes, antimicrobials and heath products during food fermentations have been well reviewed by cook.

Indigenous fermented cereal foods

Mot bacterial fermentations produce lactic acids; while yeast fermentation results in alcohol production. Many of the indigenous fermentation products of cereals are valued for the taste and aroma active components produced and are used as seasonings and condiments. A summary of flavour compounds formed in such fermented products is compiled by Chavan and Kadam. A number of fermented products utilize cereals in combination with legumes, thus improving the overall protein quality of the fermented product. Cereals are deficit in lysine, but are rich in cystine and methionine. Legumes on the other hand are rich in lysine but deficit in sulphur containing amino acids. Thus by combining cereals with legumes the overall protein quality is improved. The Chinese concept of "*fan*" (rice) and "*tsai*" (other vegetables) for a balanced and interesting diet is seen throughout the world.

8. Importance and advantages / benefits of fermented cereals:

Fermented foods contribute to about one-third of the diet worldwide. Cereals are particularly important substrates for fermented foods in all parts of the world and are staples in the Indian subcontinent, in Asia and in Africa. Fermentation causes changes in food quality indices including texture, flavour, appearance, nutrition and safety. The advantage of fermentation may induce improvement in palatability and acceptability by developing improved flavours and textures; preservation through formation of acidulants, alcohol and antibacterial compounds; enrichment of nutritive content by microbial synthesis of essential nutrients and improving digestibility of protein and carbohydrates; removal of antinutrients, natural toxicants and mycotoxins; and decreased cooking times.

Natural fermentation of cereals increases their relative nutritive value and available lysine. Bacterial fermentations involving proteolytic activity are expected to increase the biological availability of essential amino acids more than yeast fermentation, which mainly degrade carbohydrates. Starch and fiber tend to decrease during fermentation of cereals. Although, it would not be expected that fermentation would alter the mineral content of the product, the hydrolysis of chelating agents such as phytic acid during fermentation, improves the bioavailability of minerals. Changes in the vitamin content of cereals with fermentation vary according to the fermentation process, and the raw material used in the fermentation. B group vitamins generally show an increase on fermentation. Fermentation of corn meal and soybean-corn meal blend lowers flatus producing carbohydrates, trypsin inhibitor and phytates. However, fermentation of cereals with fungi, such as *Rhizopus oligosporus* has been reported to release bound trypsin inhibitor, thus increasing its activity. Fungal and lactic acid fermentations have also been reported to reduce aflatoxin B1; sometimes by opening of the lactone ring which results in complete detoxification.

FERMENTED FOODS

A. Ambil

Ingredients	**Quantity**
Sorghum flour	1000 g
Curd	500 g
Salt	To taste
Water	As per requirement

Ambil is a Maharashtra's popular fermented beverage. For the preparation of ambil the sorghum flour slurry is mixed with curd, salt and fermented overnight and then heat for 5 – 10 min, cooled and it is mixed in various proportion with species of condiments and used as beverage during meals.

B. Injera

Injera is the most popular baked product in Ethiopia. It is fermented sorghum bread with a very sour taste and is the undisputed national bread of Ethiopia. The baked product is referred to by different names depending on the locality of production in Ethiopia. It is referred to as 'bidena' in Oromigua, "taeta" in Giragiua, and "solo" in Walaytigna. According to a report by Gebrekidan and Gebrettiwat over 8% of total sorghum production in Ethiopia is used for 'Injera' production. The sorghum grains are dehulled manually or mechanically and milled to flour, which is subsequently used in the preparation of injera. The flow diagram for preparation of injera is given in **Figure 9** and various sources of flour are used for preparation and their nutrient content is given in **Table 16**.

On the basis of production procedures three types of injera are prepared.

1. **Thin injera:** This type of injera can be prepared from mixing a portion of fermented sorghum paste with three parts of water and boiling to yield a product known as 'abist' which is in turn mixed with a portion of the original fermented flour.
2. **Thick injera:** This type of injera is reddish in colour with a sweet taste, is a 'tef' paste that has undergone only minimal fermentation for 12 – 24 hrs.
3. **Komtata type injera:** This type of injera is produced from over-fermented paste and has a sour taste. The paste is baked or grilled to give a bread-like product. Yeasts are the major microorganisms involved in the fermentation of the sweet type of injera.

Sorghum flour

↓

Mix with water (4:1, w/v)

↓

Knead to form dough

↓

Mix with starter

(Fermented yellowish liquid saved from previous fermented dough)

↓

Knead

↓

Add water

↓

Ferment for 48 hr.

↓

Add water and allow standing for 1 hr.

↓

Bake on hot greased clay griddle mental till holes begin to form on top

↓

Injera

Figure 9: Flow diagram for the preparation of injera.

Table 16: Nutritional composition of injera prepared from different cereals (per 100g).

Nutrient	Sorghum	Corn	Finger millet	Barley	Wheat	Tef
Energy (Kcal)	193	185	172	167	172	162
Moisture, %	52.0	54.0	56.1	58.0	57.4	59.8
Protein, %	7.1	5.0	3.8	3.5	5.4	4.2
Fat, %	0.6	0.7	0.3	0.3	0.9	0.6
Carbohydrates, %	39.8	39.6	38.4	37.5	35.6	33.6
Fiber, %	0.9	0.7	4.0	0.9	0.9	1.7
Ash, %	0.5	0.7	1.4	0.7	0.7	1.5
Calcium, mg	10	27	169	16	28	64
Phosphorus, mg	111	120	103	128	155	129
Iron, mg	3.5	2.1	17.3	4.2	3.3	30.5
β-Carotene, μg	0	Trace	Trace	0	0	0
Thiamin, mg	0.17	0.14	0.14	0.12	0.14	0.21
Riboflavin, mg	0.08	0.06	0.01	0.05	0.09	0.07
Niacin, mg	1.7	0.7	0.2	3.0	2.4	0.8
Ascorbic acid, mg	0	2	1	0	1	1

C. Kisra

This is a thin pancake like leavened bread prepared from whole sorghum flour. It is a dietary staple food in Sudan. This fermented sorghum bread has a very sour taste. It is prepared by mixing sorghum flour with water to give a thick paste which is allowed to ferment for 12-24 hrs, following which the paste is thinned to a desirable consistency with chalked pericarp and without a sub coat gives the best sensory properties. El-Tinay reported that there was a slight increase in protein and fiber and an appreciable decrease in carbohydrate (starch and sugars) during the fermentation of Kisra. An amino acid analysis of Kisra indicated slight differences in the levels of the various amino acids. The detail procedure for preparation of Kisra is given the following flow chart (**Figure 10**).

D. Bogobe

Bogobe is a sorghum porridge prepared in Botswana from fermented and non-fermented sorghum (**Figure 11**). Fermented bogobe is a soft porridge, known as ting while the non-fermented bogobe is a thick porridge called monokwane. Information relevant to the microorganisms involved in the fermentation of bogobe and the nutritional changes, which occur during fermentation, is still scanty.

Sorghum flour

↓

Mix with water (60:40 w/v)

↓

Ferment for 12-24 hrs

↓

Add water to form thin slurry

↓

Bake on hot oiled pan

↓

Kisra

Figure 10: Flow diagram for the preparation of Kisra.

Sorghum grains

↓

Wash with water

↓

Dehull (mechanical or manual)

↓

Discard bran and grind to a coarse meal

↓

Add lukewarm water (1:1 w/v) to form slurry

↓

Allow fermenting in a closed environment for 24 hrs

↓

Cook in boiling water for 12 – 15 min

↓

Bogobe

Figure 11: Flow diagram for the preparation of bogobe from sorghum.

E. Ogi

Ogi is a porridge prepared from fermented sorghum, maize, or millet in West Africa. It is a staple of that region, and serves as a weaning food for infants. The traditional preparation of ogi involves soaking of corn kernels in water for 1 to 3 days followed by wet milling and sieving to remove bran, hull and germ. The pomace is retained on the sieve and later discarded as animal feed while the filtrate is fermented for 2 – 3 days to yield ogi, which is sour, white starchy sediment. Ogi is often marketed as a wet cake wrapped in leaves or transparent polythene bags. It is diluted to a solid content of 8 to 10% and boiled into a pap or cooked and turned into a stiff gel called "agidi" of "eko" prior to consumption.

Nutritional and microbiological studies showed that the lactic acid bacterium *Lactobacilluss plantarium*, the aerobic bacteria *Corynebacterium* and *Aerobacter*, the yeasts *Candida mycoderma, Saccharomyces cerevisiae* and *Rhodo torula* and molds *Cephalosporium, Fusarium, Aspergillus* and *Penicillium* are the major organisms responsible for the fermentation and nutritional improvement of ogi. *Lactobacillus plantarum* was the predominant organism in the fermentation responsible for lactic acid production. *Corynebacterium* hydrolysed cornstarch to organic acids while *Saccharomyces cerevisiae* and *Candida mycoderma* contributed to flavour development.

Steeping, milling and sieving are the processing steps during which considerable nutrient losses take place. Much of the protein in cereal grains is located in the testa and germ, which are usually sifted off during processing. A protein enriched ogi containing 10% soybeans flour was developed by the Federal Institute of Industrial Research Oshodi (FIIRO) Lagos, Nigeria. The development of an ogi product (*dogik*) having therapeutic properties on the basis of its ability to control diarrhea among infants. This finding is of great relevance since ogi is used as a popular weaning food for children in African countries. Odunfa *et al.* (1994) investigated the possibility of improving the limiting lysine level in ogi. Fifty mutants from *Lactobacillus plantarum* and seven mutants from yeast strain were selected from thialysine-resistant cultures capable of overproducing lysine and analyzed for lysine production. Up to 12 fold increase in lysine production was observed from *Lactobacillus plantarum* and 3 – 4 fold increase from yeasts was observed. Utilization of the mutants as starter cultures resulted in a three-fold increase in the lysine content of ogi. The use of high lysine corn for improving the nutritional value of ogi is the best way. Dehydration of ogi by drum or tray drying has been shown to prolong its shelf life. Drum drying was however reported to destroy heat sensitive nutrients in ogi. Some scientist reported that an appreciable loss in the available lysine content of ogi as a result of drum drying. Bacteriocin-producing *Lactobacillus* isolated to be active against common food-borne pathogens including *Salmonella*. This bacteriocin also improved the shelf life of "jellied" ogi extending it by 10 days.

F. Masa

This is fermented bread made in Nigeria. A similar batter is used, though thicker in consistency and mixed with onion, pepper, and salt before fermentation. It is fried in oil on both sides as small individual hemispheres.

G. Gahlet

It is a small thick leavened cakes made in Burkina Faso. The weaning food, nasha is made from the same kind of flour and water mixture as Kisra, but is allowed to ferment for 2 days or longer. It is then added to about half the volume of boiling water, and cooked until it thickens.

H. Quitta

This is unleavened bread used to only a small extent in Ethiopia.

I. Waina

This is unleavened bread made in Nigeria; it contains hot pepper and spices, and a substantial amount of groundnut oil. These ingredients are mixed to form dough, which is cut into flat discus and fried in oil.

J. Stiff porridges

The stiff porridge is made by adding flour steadily to boiling water, stirring all the time to avoid lumps, then cooking. The product should be stiff enough to mould without cracks, it should be neither sticky nor crumbly. Names of porridges according to country and region; *Ugali* in East Africa, *Genfo* in Ethiopia, *Guma* in Sudan, and *Tuwo* in Nigeria.

K. Tortillas

The sorghum grains are cooked in limewater in the first stage in Latin America. Prepare mesa by mixing lime solution (0.5 percent) and sorghum grain in 3:1 proportion and cooking for 3 to 10 minutes at the boiling point. Steep sorghum grains for at least 4 hours. Prepare balls from the mesa and press them into circles of about 15 cm diameter and 0.5 cm thickness. Cook the tortillas on a grill or a traditional clay comale. During cooking turn the tortilla once to brown it lightly on both sides. Leave the cooked tortillas on the floor to cool a little, and then keep them in a container lined with a cloth to cover.

L. Couscous

This type of product is mostly prepared in West Africa. Wet the finely ground sorghum flour with cold water and knead it until flour particles agglomerate. Force the mixture through a fine screen (1.5 mm mesh). Place the grains in a perforated pot fitted over another pot containing boiling water. Put a cloth seal at the joint between the two pots. Heat the lower pot to steam the grains above for about 15 minutes. They will form a single large chunk. Take out the chunk, break it into small aggregates and transfer them back to steam for another 15 minutes. Remove the chunk, break it into aggregates and sift them through a sieve (2.5 mm, mesh). Dry and store for future use. To prepare couscous for serving, sprinkle cool water on the aggregates. Mix thoroughly with fingers. Mix the grains with ground baobab leaf powder or okra powder and other ingredients such as peanut paste, etc. and give it a final steaming for 15 minutes. Allow it to cool slowly. Serve with sauce or milk, or dry it and use as a convenience food.

M. Fura

Fura product is mostly prepared from sorghum and used as a main food in Nigeria.

Ingredients	Quantity
Sorghum flour	200 g
Hot spices	2 tea spoons
Yoghurt or sour milk	100 ml
Water	300 ml approximately

First mix flour, water and spices together. Prepare small round balls (2-3 cm in diameter). Drop them into boiling water and cook for 30 minutes. Pound cooked balls with water and spices until a smooth, elastic and cohesive lump is formed. Again prepare small balls; rolling between the palms of the hand or on a wooden board dusted with dry flour. Serve as it is or with nono, yoghurt or sour milk, as a snack.

Sorghum Beers

In several African countries sorghum malt is fermented into beer. The beer making process may depend upon local practices and types of cultivars used. Use of sorghum as a brewing grain is gaining popularity not only in other countries of African but also places like Mexico due to rising import cost of barley malt. Some researchers carried out trials on use of sorghum malt in making layer beer in Botswana. A three stage decoction system of mashing was used and the product was acceptable the consumers. A popular process is being used for making Nigerian sorghum beer in a traditional way. European types of beers were investigated by studying bottom fermentation brewing with Nigerian sorghum malt, sorghum barley beers scored identically. Recently pilot experiment was done successfully for processing sorghum malt to produce lager beer using Nigerian sorghum. In recent year some scientists in India prepared pure sorghum and sorghum-barley malt based beer, which were acceptable to the consumer panel. Ogundiwin and Tehinse conducted fermentation test on malted sorghum wort in the traditional Nigerian pito drink. Effect of adding enzymes to a mixture of malt and sorghum during brewing process on taste, foam stability and chill sensitivity qualities of beer was studied and it was observed that here was increase in soluble carbohydrate and nitrogen, decrease in wort viscosity, higher filtration rate and increased yield. The sorghum grain is without husk unlike barley and hence it causes wort filtration problems.

Karif beer has always as a good and acts as a source of B vitamins to Banu people. It has been reported that beer has calorific value of 400 – 500 Kcal/Lit and thiamin 1 –2 mg/lit, Novellie (1977) reported high vitamin contents of sorghum beer. Heerden analyzed commercially brewed sorghum beer for thiamin, riboflavin, nicotinic acid and protein and reported that sorghum beer had higher thiamin, riboflavin and mineral content than barley beers. Whole sorghum beers were more nutritive than sorghum beer with maize grits as adjuncts. A new sorghum beer product, known as "Long life" is being produced in South Africa, pasteurized and aseptically packaged in tetra pack. Okrafor and Aniche have developed a hop substitute for sorghum beer using a West African herb *Gongronema lactifolium* locally known as "Utazi".

Merissa beer is prepared from sorghum. The sorghum grains are soaked, malted for 2 days and milled to coarse grain. Ajeen, fermented dough is obtained by mixing sorghum flour with water and incubating for 36 hrs. The ajeen is cooked in hollow steel containing until it turns dark brown (soorji). The soorji is then mixed with 5% malt flour, 5% good Merissa and water and allowed to ferment for 4 to 5 days (deboba). A cooked sorghum paste, futtara, is obtained by mixing cooked greenish yellow and brown pastes. Malt flour is mixed with the futtara and placed underneath the doboba surface as a solid mass. Fermentation and liquefication proceed for 8 to 10 hrs. The mass is strained to yield merissa. The principal microorganisms in Merissa fermentation are lactic acid bacteria and *Saccharomyces spp.* of yeast. The detail procedure is given in **Figure 12**.

Conversion time is the saccharification rate during mashing process. Shorter the conversion time better is the brewing quality of the malt. CSH – 9 and HES – 4 required 7 – 10 min of conversion time and this is considered good. SPV 297 and M 35 –1 required conversion time 10 – 15 min, which is also in acceptable range. The specific gravity of sorghum malt extracts was slightly lower as compared to the literature values for barley malt extracts. The specific gravity of wort from HFS – 4 and M 35 –1 was comparable. HES – 4 and M 35 –1 had higher extract yields as compared to the other two cultivars. Iodine reaction of all wort samples was yellowish indicating absence of starch in the wort and hence sorghum worts were comparable. Colour of the wort from sorghum malt was darker as compared to barley wort. Wort viscosity of M 35 – 1 was slightly higher as compared to the wort viscosity of other cultivars. However, the viscosity of sorghum wort was comparable to barley wort (**Table 17**).

Total nitrogen content of beer wort of HES – 4 was higher as compared to the values for other three sorghum cultivars. However, mean total nitrogen content to the values for other three sorghum cultivars was comparable to the total nitrogen content of barley wort. Free amino nitrogen (FAN) content of sorghum worts was slightly lower as compared to the barley wort but the varietal differences were also observed. The extract of FAN formation is a function of proteolytic activity in the malt as well as action of carboxy peptidase action during mashing at lower temperature. A balance must be struck between breakdown of proteins to amino acids for yeast nutrition and preservation of high molecular weight proteins, which enhance the foaming properties of the beer. Excessive proteolytic activity may cause hydrolysis of high degree of the large molecular weight.

Neutral polypeptides result into poor foaming and head retention in beer. The third condition with respect to the removal of proteins combine with polyphenols from acid protein fraction in beer, which contributes to haze formation in beer, hence its early removal enhances the stability of beer. Hence removal of basic proteins in wort is essential. However, if the basic protein is insoluble by fixing to the spent grains, it may act as a wort clarifier and remove troublesome polyphenols from the wort.

Total sugars and reducing sugars of the sorghum wort and barley wort were comparable. The varietal differences were negligible for sugars. The values of total sugars and reducing sugars were comparable to those reported by earlier workers. Some scientists reported that lower reducing sugars in wort and hence they used 30 percent sucrose as adjunct in the wort prior to fermentation. Several studies have been conducted on barley malt mashing and the resultant fermentable sugar content due to enzymatic degradation of starch during mashing has been considered as the most important biochemical reaction. Acidity and pH of sorghum wort was comparable to that of barley wort and this value is in acceptable range required for the beer fermentation.

After the wort is cooled it is pitched with required quantity of yeast and then immediately the fermentation begins. The first sign of fermentation is the production of gas bubbles, which collects as forth. The fermentation is normally completed in 7 days. The growth of the yeast in fermenting wort causes several biochemical changes particularly it respect to specific gravity, extract content, pH and sugars which are reduced in quantity, where as there is increase in alcohol production. There is significant reduction in specific gravity of wort during beer fermentation and the reduction is proportionate to the time of fermentation. The differences in the rate of specific gravity reduction were negligible due to cultivars in sorghum wort. There was reduction in extract content of the wort and the rate of change of extract content was similar to that of reduction in specific gravity. The reduction in extract content was from 9.24 to 2.52 in 7 days of fermentation of wort from HES – 4 sorghum. During 7 days of fermentation the wort pH was reduced in all the samples. The change was proportionate for all the sorghum cultivars and comparable to barley wort.

Reducing sugar was found to be reducing proportionately and at almost constant rate as fermentation proceeded. A slight variation in reduction was observed due to cultivar but overall the rate of and the extent of reduction in reducing sugar constant of sorghum wort was comparable to barley wort. The yeast utilized sugars to produce alcohol as the end product. The rate of reducing sugar utilization by yeast was high during first 5 days of fermentation. The rate of utilization of reducing sugar decreased thereafter.

The formation of alcohol during the fermentation of beer wort is a function of yeast growth. The rate of increase in alcohol content was proportional to the yeast growth and consumption of sugars. The rate of alcohol production was constant upto 5 days and then it decreased slightly. Sorghum varietal differences in the rate of alcohol production during fermentation were very negligible and were comparable to barley wort fermentation. The choice of yeast, the condition of yeast at the time of pitching, the amount of yeast added, the distribution of yeast and type of fermenting vessel, aeration, wort composition and pH, fermentation temperature are some of the main factors influencing fermentation performance and beer quality.

Specific gravity of beer from sorghum cultivars ranged from 1.0092 to 1.00985. The apparent extract of sorghum beer was not affected by cultivar and was comparable. Real extract content of beer from CSH – 9 was higher than beer of HES – 4 and M 35 –1 but it was comparable with control barley beer (**Table 18**). Attenuation is the degree of fermentation in beer based on loss in the specific gravity of the wort. Apparent attenuation of sorghum beers was comparable for all the four cultivars. However, sorghum beer had slightly higher apparent degree of fermentation as compared to control barley. The variation in samples for attenuation could be attributed to basic differences in wort composition with respect to available nutrients for the growth of the yeast. Real attenuation for CSH – 9 and SPV – 297 were comparable with barley beer whereas M 35 – 1 and HES – 4 had higher percent real attenuation. A high attenuation is the result to wort composition. Attenuation limit can be increased by addition of external enzyme sources.

The scores obtained from various sensory characters like colour, clarity, body, aroma, taste and overall acceptability were statistically compared after transmission of sources to square roots. Colour of the beer prepared from HES – 4 and M 35 –1 sorghum malt was significantly superior over control barley malt beer as well as beer prepared from other sorghum cultivars. However, commercial beer sample had significantly superior beer colour. The colour of the beer is largely due to the melanoidins and caramel present in the malt and adjuncts used. Further caramelization can occur during wort boiling and major adjustments to the colour of beer can be made by the addition of caramel. Hence varietal impact on beer colour is of very little value in the control of beer quality (**Table 19**).

Body of beer refers to fullness given by the beer sip in the mouth. Perhaps this is related to beer viscosity as felt in the mouth. The body of the beer sample of M 35 – 1 was comparable with commercial beer sample. All the experimental beer samples had acceptable beer aroma. The principle higher alcohols found in beer are 3-methyl butanol, 2-methyl butanol, propanol and phenethyl alcohol, which are responsible for beer aroma. Sensory evaluation of taste of experimental beer samples indicated that all the samples were in acceptable range. From all these experimental information it can be concluded that sorghum can be very well used for the production of beer.

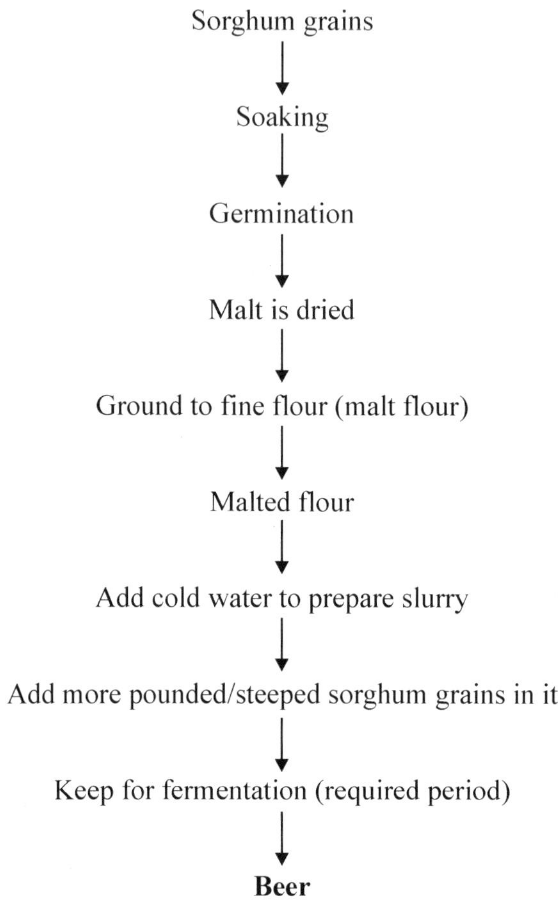

Figure 12: Preparation of sorghum beer.

Enzymes in the malt begin to saccharify the starch. Bacteria then act on a part of the sugar to form lactic acid and some B vitamins. This sweet wort may be drunk after less than one-day fermentation. It contains more B vitamins and negligible amount alcohol. After the first day, the mixture is heated and various fungal spores germinate and lead to the production of hydrolytic enzymes. A little millet or sorghum may be specially germinated, then dried as a yeast source. The yeasts multiply and fermentation develops, with the production of alcohol, more B vitamins, and new proteins, while the bacteria continue to develop lactic acid. The brew may be drunk after 4 – 5 days, and in some areas more malted grain may be added to it a day before it is used.

Table 17: Properties of sorghum malt for brewing

Genotype	Conversion time, min.	Speed of filtration	Degree of clarity	Colour of malt EBC units
CSH – 9	7 - 10	Slow	Slightly hazy	1.85
SPV – 297	10 – 15	Slow	Slightly hazy	1.60
M 35 – 1	10 – 15	Slow	Clear	1.55
HES – 4	7 – 10	Slow	Clear	1.80

Table 18: Comparative evaluation of worts from sorghum and barley malt.

Genotype	SP. gravity	Original extract optical density	Iodine reaction	Colour EBC units	Viscosity Cp
CSH – 9	1.03360	8.444	Yellowish	9.20	2.42
SPV – 297	1.03435	8.620	Yellowish	9.00	2.61
M 35 – 1	1.03602	9.060	Yellowish	8.00	2.85
HES – 4	1.03690	9.240	Yellowish	9.40	2.35
Barley	1.04160	10.380	Yellowish	7.80	2.45

Table 19: Sensory evaluation of sorghum beer as affected by cultivars*.

Genotype	Colour	Clarity	Body	Aroma	Taste	Overall acceptability
CSH – 9	2.64	2.51	2.55	2.53	2.56	2.66
SPV – 297	2.65	2.51	2.60	2.64	2.60	2.69
M 35 –1	2.67	2.43	2.77	2.74	2.65	2.76
HES – 4	2.75	2.52	2.76	2.72	2.61	2.74
Barley	2.59	2.75	2.87	2.84	2.75	2.84
Commercial	2.82	2.98	2.80	2.97	2.94	2.94
S.E. ±	0.032	0.031	0.025	0.024	0.022	0.021
CD at 5%	0.089	0.087	0.070	0.067	0.061	0.060

*Values are means of square root transformed scores.

Bhakari / Roti (unleavened bread)

This is unleavened bread widely used in India. Whole sorghum is finely ground, a little salt is added, and it is mixed with hot water little by little to make separate balls of dough. These are pressed by hand in to flat discs, and baked on both sides in a shallow pan. Cooking time depends upon the thickness of the disc. *Roti* should remain soft a full day and there are genotypic differences in this character.

Dough quality: flour that can give good, cohesive dough generally produces an acceptable *roti*. Dough cohesiveness was evaluated with the Instron food-testing instrument using a back extrusion cell. Cohesive doughs yielded a characteristic back extruction profile requiring less force for deformation. Poor quality dough that is less cohesive gets compressed into the cell without extrusion extracts of flour on dough stickiness **(Plate 1)**.

The relationship between the physicochemical characters and roti quality indicated that quantity of water-soluble proteins; amylose and sugars jointly influence the roti quality.

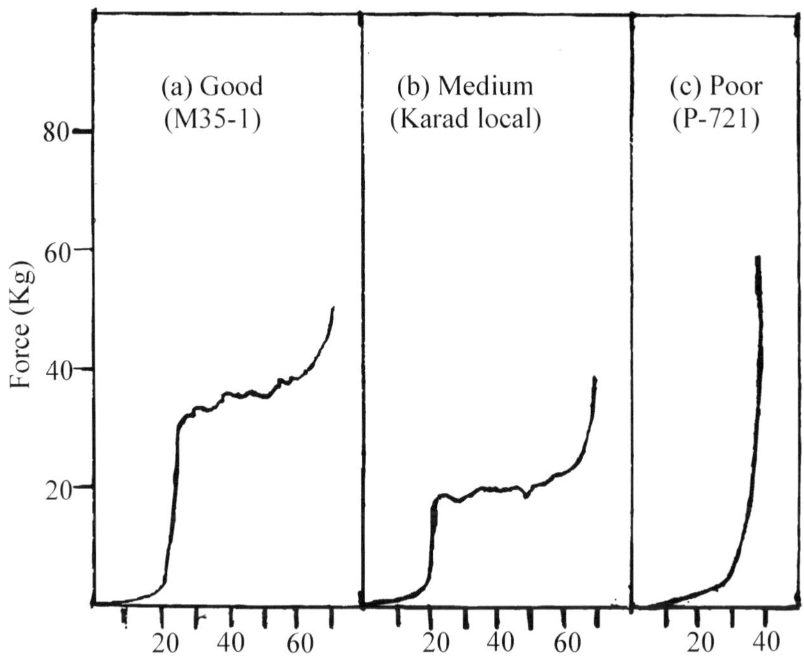

Plate No. 1: Force-distance curves for good, medium and poor quality dough's of sorghum flour (Instron).

The most common use of sorghum in India is in the form of flour or grit, which is transformed into dough and baked on an open hot pan (*Tawa*) as falt unleavened bread i.e. *roti* or *bhakri*. *Bhakri* is the traditional type round pancake made by the traditional method. A large number of diabetic patients are regularly consuming sorghum bhakari with bitter gourd or fenugreek *subji* through every day meals. However, the bhakari consumption has been restricted owing to its course nature and palatability. For obtaining good quality bhakari flour, a 60-mesh sieve could be used for separation of the seed coat from the whole flour.

However, the whole meal flour is used for the special bhakari making intended for the diabetics and other patients.

Sorghum hybrids, in general are rated low for consumption as compared to local or varieties/cultivars because of numerous consumer complaints. The poor *bhakari* making quality is exemplified by low water-holding capacity of the flour, difficulty in kneading, cracking and low extensibility of dough, crumbling during baking and low water retention.

Procedure for *bhakari* preparation

Sieve sorghum flour using 60-mesh sieve. Add small quantity of salt if required, hot water and knead into smooth soft dough. Make round small ball approximately 50 or 100 g weight and spread it into a round sheet using rolling pin or by pressing with hand palms or using modern roti making machine. Bake it on pre-heated pan (*Tawa*) or iron sheet. Apply immediately water on the top of bhakari while baking. After that change the up side down and partially bake it on pan/Tawa or iron sheet. Finally place it directly on hot coke or gas flame and bake till we get good puffed bhakari. Such type of bhakari serve immediately hot i.e. fresh. The flow sheet for the preparation of *bhakari/roti* is given in **Figure 13**. The nutritional composition of *bhakari* is given in **Table 60.**

Preservation of *bhakari*

The sorghum *bhakari* can be preserved in LDPE bags (200 gauge) or laminated pouches by vacuumization packaging technique. This technique showed that *bhakari* remains soft and retain their palatability upto two days under ambient conditions. The *bhakari* making and preservation technique could be used on commercial scale so as to make it easily available in the market for ready consumption.

Extension shelf life of *bhakari*

Traditionally, the sorghum *bhakari* are consumed as fresh i.e. within eight hours after preparation. Addition of 2 – 3 percent phosphatised starch or 2 percent guar gum in the sorghum flour significantly improved the *bhakari* taste and palatability by retaining maximum amount of moisture. Packaging *bhakari/roti* in laminate pouches after vacuumization found to increase the shelf life by 2 days.

Nutritional and *bhakari* making qualities of some improved cultivars of grain sorghum

The physical properties, nutritional composition, digestibilities of protein and starch and *bhakari* making quality of some improved sorghum cultivars have been investigated. Both protein and starch digestibilities varied markedly among the cultivars studied. The cultivars M 35 – 1, SPV – 86 and CSH – 8R exhibited superior sensory properties of *bhakari* over other cultivars (**Tables 20, 21, 22**

and 23). The grain weight, soluble proteins and soluble sugars were found to be positively correlated with *bhakari* quality. These results indicated that certain hybrids are comparable in both nutritional and bhakari quality with varieties. It is suggested that while breeding varieties for higher yields, attention needs to be paid for obtaining higher grain weight, soluble proteins and soluble sugars in the newly developed high yielding cultivars.

Sorghum Grains
↓
Cleaning
↓
Grinding
↓
Sieve with 60-mesh sieve (for fine flour if required)
↓
Flour
↓
Add salt and hot water
↓
Knead into smooth dough
↓
Make small balls (50 or 100g)
↓
Roll dough into round pan/iron sheet or
use *bhakari*-making machine to get uniform thickness of *bhakari*
↓
Bake on hot iron pan/Tawa
↓
Apply water on top of *bhakari*
↓
Remove it by flat iron spoon after 2-3 min.
↓
Make upside down and bake sufficiently to form semi hard body
↓
Finally bake it on hot coke or on gas flame
↓
Bhakari puffs into two layers (upper thin and lower thick)
↓
Serve it as hot or cool to room temperature
↓
Store in moisture proof paper or container
↓
Vacuumise it, seal and store for 3 – 4 days
(It remains in good condition)

Figure 13: Flow sheet for preparation of *bhakari* from sorghum and other millets.

Table 20: Physical properties and nutritional quality of different cultivars of grain sorghum.

Cultivar	Grain wt. (g/100 seeds)	Grain hardness (kg)	Protein (% N X 6.25) (%)	Soluble proteins (%)	Starch (%)	Reducing sugars (%)	Total sugars (%)	IVPD* (%)	IVSD* (mg maltose/ g/2 hr)
Varieties									
M 35 – 1	3.89	6.4	11.90	1.33	76.23	0.14	2.00	69.07	181
CSV – 4	3.52	7.5	13.20	1.30	73.50	0.07	0.86	60.39	200
SPV–504	4.17	5.9	10.50	1.18	77.24	0.10	0.90	72.86	160
SPV–351	2.50	8.5	11.77	1.07	75.42	0.08	0.78	78.91	135
SPV – 86	3.81	6.4	10.12	1.15	76.72	0.08	0.87	73.12	168
Hybrids									
CSH – 1	3.34	6.0	9.62	1.15	78.32	0.10	0.81	59.41	150
CSH – 5	3.42	6.8	11.38	1.08	76.81	0.10	1.00	75.07	140
CSH – 6	2.91	6.2	11.38	1.08	76.81	0.10	1.00	67.41	184
CSH – 9	3.53	5.2	10.58	1.19	76.93	0.11	1.24	68.67	140
CSH – 8R	3.19	5.8	10.77	0.98	76.55	0.14	0.96	60.63	166

*IVPD = In-vitro protein digestibility; IVSD = In-vitro starch digestibility.

Table 21: *Bhakari* making quality of different cultivars of grain sorghum.

Cultivar	Water uptake (ml/50 g of flour)	Dough rollability (cm)	Percent moisture loss in *bhakari* after storage (hrs)			
			3	6	12	24
Varieties						
M 35 – 1	50.7	21.3	0.88	4.98	7.99	17.21
CSV – 4	58.0	22.5	1.53	4.19	7.85	15.02
SPV – 504	53.0	22.9	1.06	5.01	6.80	15.56
SPV – 351	53.0	21.8	2.42	4.97	8.17	16.61
SPV – 86	56.0	22.8	2.07	4.28	7.88	13.89
Hybrids						
CSH – 1	51.7	22.2	1.92	4.49	7.93	15.33
CSH – 5	57.7	22.7	2.73	5.19	8.98	16.69
CSH – 6	53.7	24.5	1.82	5.04	9.88	16.48
CSH – 9	52.7	24.0	1.58	4.41	8.11	15.67
CSH – 8R	52.0	23.3	2.88	5.17	8.64	15.85

Table 22: Sensory properties of sorghum *bhakari**.

Cultivar	Colour and appearance	Texture	Flavour	Sweetness	Mean
Varieties					
M 35 – 1	9.00	8.75	8.00	7.94	8.43
CSV – 4	7.25	7.06	7.50	7.50	7.35
SPV – 504	7.25	7.63	7.69	7.44	7.50
SPV – 351	6.63	6.75	6.38	6.75	6.65
SPV – 86	8.25	7.75	7.63	7.75	7.88
Hybrids					
CSH – 1	3.75	5.88	6.25	6.13	5.53
CSH – 5	6.75	7.38	6.88	7.13	7.05
CSH – 6	7.38	7.25	7.00	7.75	7.38
CSH – 9	7.25	7.75	7.50	7.25	7.48
CSH – 8R	7.75	8.00	7.50	7.75	7.78
S.E. ±	0.28	0.29	0.28	0.31	0.23
CD at 5%	0.79	0.81	0.80	0.89	0.68

*Sensory score out of 10 point hedonic scale using 5 semi-trained judges

Table 23: Correlation coefficient between physico-chemical characters and sensory properties of sorghum *bhakari*.

Character	Colour and appearance	Texture	Flavour	Sweetness	Acceptability
Grain weight	+ 0.722	+ 0.367	+ 0.314	+ 0.476	+ 0.443
Grain hardness	- 0.387	- 0.326	- 0.029	- 0.309	- 0.199
Soluble sugars	+ 0.368	+ 0.353	+ 0.292	+ 0.597	+ 0.398
Soluble proteins	+ 0.484	+ 0.219	+ 0.235	+ 0.232	+ 0.277

Effect of grain conditioning on physico-chemical properties of sorghum *bhakari*

Sorghum grains conditioned with 10% extra moisture before traditional milling exhibited 9 to 11 percent removal of bran during household sieving. The resultant flour exhibited improved water absorption during dough preparation and weight of bhakari during storage. The boiling water for dough preparation accentuated this effect. The conditioning and removal of bran did not influence the nutrient composition of bhakari markedly except significant lowering the fiber content.

The results indicated that eating quality and storability of sorghum bhakari could be improved by simply conditioning treatment to grains prior to milling without adversely affecting the nutritional composition. Effect of processing on the nutritional and other parameters are given in **Tables 24, 25, 26**, and **27**.

Table 24: Effect of conditioning the grains on the removal of bran.

Conditioning moisture (%)	Percent bran removed*		
	M 35 – 1	CSH - 5	CSH – 8R
10 (control)	5.0	4.4	4.4
15 (5% extra)	8.0	8.4	7.4
20 (10% extra)	11.6	10.4	9.6

*Means of duplicate experiments

Table 25: Effect of conditioning the grains on water absorption by resultant flour.

Conditioning moisture (%)	Water temperature for dough preparation (°C)	Percent water absorption*		
		M 35 -1	CSH - 5	CSH –8R
10 (control)	25	82	86	82
20 (10% extra)	25	94	92	92
10 (control)	50	86	90	86
20 (10% extra)	50	94	94	94
10 (control)	Boiling water	90	96	88
20 (10% extra)	Boiling water	104	98	100

*Means of duplicate experiments.

Table 26: Loss in weight of *bhakari* prepared by using cold (25°C) and boiling water during storage.

Cultivar	Conditioning moisture level (%)	Dry weight of bhakari (g) during storage (hrs)									
		Cold water					Boiling water				
		0	3	6	12	24	0	3	6	12	24
M 35-1	10 (control)	78.00	74.75 (4.2)	74.25 (4.8)	73.50 (5.8)	72.75 (6.7)	81.00	76.00 (6.2)	75.00 (7.4)	73.50 (9.3)	70.50 (12.4)
	20 (10% extra)	84.00	79.75 (5.1)	79.00 (6.0)	78.00 (7.1)	75.90 (9.6)	88.00	82.75 (6.0)	81.75 (7.1)	80.00 (9.1)	78.50 (11.9)
CSH-5	10 (control)	80.00	76.00 (5.0)	75.50 (5.6)	74.50 (6.9)	73.00 (8.8)	84.00	79.00 (6.9)	78.00 (7.1)	77.00 (8.3)	76.00 (9.5)

Cultivar	Conditioning moisture level (%)	Dry weight of bhakari (g) during storage (hrs)									
		Cold water					Boiling water				
		0	3	6	12	24	0	3	6	12	24
	20 (10% extra)	83.00	79.75 (4.0)	79.00 (4.8)	78.25 (5.7)	76.75 (8.1)	85.00	81.00 (4.7)	80.00 (5.9)	79.00 (7.1)	78.00 (8.2)
CSH-8R	10 (control)	78.00	74.50 (4.5)	74.00 (5.1)	73.50 (5.8)	72.00 (7.7)	80.00	77.00 (3.8)	76.00 (5.0)	75.00 (6.3)	74.00 (7.6)
	20 (10% extra)	83.00	79.00 (4.8)	78.50 (5.4)	78.00 (6.0)	75.50 (9.0)	86.00	82.00 (4.7)	81.00 (5.8)	79.75 (7.3)	77.25 (10.2)

*Figures in parentheses indicate percent weight loss.

Table 27: Proximate composition of *bhakari* prepared from different resultant flours.

Cultivar	Conditioning moisture level (%)	Protein (%)	Crude fat (%)	Crude fiber (%)	Ash (%)	N-Free extract (%)
M 35 – 1	10 (Control)	10.55	3.51	2.05	1.77	82.12
	20 (10% extra)	10.41	3.27	0.80	1.58	83.94
CSH – 5	10 (Control)	9.02	3.61	1.95	1.58	83.84
	20 (10% extra)	8.95	2.76	0.89	1.55	85.85
CSH –8R	10 (Control)	8.94	3.26	1.95	1.80	84.05
	20 (10% extra)	8.89	3.17	0.72	1.72	85.50

Effect of incorporation of wheat, maize, black gram, and defatted soybean flour in sorghum flour on physical and sensory properties of *bhakari*:

Mixing of wheat, maize, black gram and defatted soybean flour to sorghum flour resulted in the decrease in water absorption by the flour during dough making, decreased the fresh weight and increase water loss in *bhakari* during storage. However, the differences in the fresh weight of *bhakari* were not of significant order. Wheat can be mixed to the extent of 10 percent while maize, black gram and defatted soybean can be mixed to the extent of 20 per cent with sorghum to produce acceptable *bhakari* with improved protein and ash contents. An increase in protein content in such *bhakari* ranged from 8.15 to 17.66 per cent when black gram and defatted sorghum were mixed. The detail data of the experiment is given in **Tables 28, 29, 30, 31** and **32.**

Table 28: Sensory properties of *bhakari* prepared from sorghum : wheat flour blends

Flour blend, % Sorghum:Wheat	Colour & appearance	Flavour	Texture and body	Sweetness	Overall acceptability
100:O	7.75	8.00	8.25	8.25	8.06
90:10	7.00	7.50	7.61	7.25	7.34
80:20	6.75	7.27	7.25	7.25	7.13
70:30	6.50	7.25	7.25	7.25	7.06
S.E. ±	0.24	0.23	0.28	0.20	0.24
CD at 5%	0.77	0.76	0.89	0.66	0.77

Table 29: Sensory properties of *bhakari* prepared from sorghum : maize flour blends.

Flour blend, % Sorghum:Maize	Colour & appearance	Flavour	Texture and body	Sweetness	Overall acceptability
100:O	7.75	8.00	8.25	8.25	8.06
90:10	8.50	7.25	7.00	8.00	7.68
80:20	7.90	7.20	7.00	7.50	7.40
70:30	7.00	6.75	6.75	7.25	6.94
S.E. ±	0.18	0.17	0.24	0.24	0.21
CD at 5%	0.57	0.54	0.77	0.77	0.66

Table 30: Sensory properties of *bhakari* prepared from sorghum : black gram flour blends.

Flour blend, % Sorghum : Black gram	Colour & appearance	Flavour	Texture and body	Sweetness	Overall acceptability
100:O	8.50	8.25	8.25	8.50	8.38
90:10	8.25	8.00	7.50	7.75	7.88
80:20	7.75	8.00	7.50	7.25	7.63
70:30	7.50	7.50	7.25	6.50	7.19
S.E. ±	0.31	0.21	0.34	0.24	0.28
CD at 5%	1.02	0.68	1.12	0.77	0.90

Table 31: Sensory properties of *bhakari* prepared from sorghum : soybean flour blends.

Flour blend, % Sorghum: Soybean	Colour & appearance	Flavour	Texture and body	Sweetness	Overall acceptability
100:O	8.50	8.25	8.25	8.50	8.38
90:10	8.00	7.75	8.25	8.25	8.06
80:20	8.75	7.75	8.00	8.25	7.69
70:30	6.50	7.00	6.50	7.00	6.75
S.E. ±	0.36	0.27	0.20	0.33	0.29
CD at 5%	1.16	0.87	0.66	1.05	0.94

Table 32: Proximate composition of *bhakari* prepared from various composite flours.

Sample	Protein (%)	Crude fat (%)	Crude fiber (%)	Ash (%)	N-free extract (%)
Sorghum (CSH-5, 100%)	8.15	3.31	0.80	1.55	86.19
Flour blends					
Sorghum : Wheat					
0:100	14.07	2.46	1.06	1.70	80.71
90:10	10.26	2.70	1.40	1.52	84.12
80:20	11.49	2.82	1.33	1.54	82.82
70:30	12.64	2.95	1.28	1.60	81.53
Sorghum:Maize					
0:100	11.96	4.30	0.75	2.58	80.41
90:10	8.49	3.00	0.76	1.70	86.05
80:20	9.27	3.32	0.72	1.88	84.81
70:30	9.68	3.60	0.70	2.02	84.00
Sorghum:Blackgram					
0:100	22.09	2.03	0.10	3.32	72.46
90:10	10.43	2.50	0.75	1.85	84.47
80:20	12.00	2.33	0.70	2.03	82.94
70:30	12.73	2.19	0.65	2.32	82.11

Sample	Protein (%)	Crude fat (%)	Crude fiber (%)	Ash (%)	N-free extract (%)
Sorghum:Soybean					
0:100	53.46	4.20	4.57	5.82	31.95
90:10	12.68	2.94	1.60	2.24	80.54
80:20	17.66	3.00	2.35	2.85	74.14
70:30	21.79	3.25	3.21	3.34	68.41

Utilization of groundnut flour in preparation of sorghum *bhakari/ roti*

Sorghum is consumed as one of the staple food grains in India. It is mostly consumed in the form of unleavened traditional products, *bhakari* and *sankati*. However, these products are low in protein content and exhibit poor sensory properties, which limit the acceptability of sorghum, particularly, that of high yielding hybrids by the consumers. The defatted oilseed flours contain 40 to 60% proteins and thus form a potential protein source to the vegetarian diets. Such flours can be supplemented with low-protein cereals meals to facilitate their utilization in human foods and also to improve the protein content of cereal-based foods. An attempt has been therefore, made to supplement sorghum meal with 10 to 30% defatted groundnut flour in the preparation of sorghum *bhakari* and to study its acceptability.

The defatted groundnut flour (prepared in the laboratory by solvent extraction of full-fat meal) was blended with sorghum flour (cv. CSH – 9) in the proportion of 0, 10, 20, and 30% (w/w). The water uptake by the composite flour, dough rollability and moisture loss from *bhakari* during storage were studied. Fresh *bhakari* was evaluated by panel of judges for various sensory properties such as colour and appearance, flavour, texture, and sweetness using 10-point hedonic scale. Average of five judges were used to determine the acceptability of the product.

Mixing of defatted groundnut flour with sorghum flour to the extent of 20 to 30% significantly decreased the water uptake by the flour during dough preparation. However, the rollability of dough and the percent water loss from the *bhakari* during storage upto 12 hrs were not significantly affected. Blending of groundnut flour to the extent of 30% significantly improved colour and appearance, texture and sweetness of sorghum *bhakari*. The flavour of composite *bhakari* was not significantly affected by addition of defatted groundnut flour. The results indicated that mixing of groundnut flour significantly improved the overall sensory properties and acceptability of sorghum *bhakari*.

Mixing of 30% of defatted groundnut flour containing about 40% proteins with 70g of sorghum flour containing 10% proteins, would give composite flour with about 19% protein content. Moreover, the composite *bhakari* is more acceptable than the product prepared from 100% sorghum flour (**Table 33**).

Table 33: Physical and sensory properties of *bhakari* prepared from sorghum:groundnut composite flour.

Flour blend (w/w) Sorghum: Groundnut	Water uptake (ml/50g of flour)	Dough rollability (cm)	Water loss during storage 12 hrs, (%)	Sensory qualities of sorghum bhakari				
				Colour & appearance	Flavour	Texture	Sweetness	Overall mean
100:0	50.00	20.00	11.62	8.00	7.80	7.80	8.00	7.90
90:10	48.50	20.15	11.91	7.40	7.20	7.40	7.80	7.45
80:20	42.00	20.30	11.67	8.60	7.80	8.40	8.40	8.30
70:30	41.00	20.00	12.90	8.80	8.20	8.80	9.00	8.70
CD at 5%	1.72	N.S.	N.S.	0.52	0.62	0.53	0.38	0.56

N.S. = Non-significant.

Nutritional improvement and utilization of grain sorghum by fermentation

A natural lactic fermentation of ground grain sorghum produced significant increase ($P < 0.001$) in available lysine/leucine, isoleucine, and methionine. The protein quality, expressed as relative nutritive value, increased significantly ($P <0.001$) as a result of fermentation. Niacin and thiamine increased significantly ($P <0.001$) and riboflavin also increased ($P < 0.05$) during the fermentation (**Table 34**). Protein and carbohydrate appeared to be more available after the fermentation than before the fermentation. Although the fermentation produced increased availability of nutrients, it did not produce change in the proximate analyses.

The relative nutritive value (% RNV) of sorghum fermented at 25°C and 35°C increased significantly ($P < 0.05$) over the% RNV of the control. During the consecutive 7-day fermentation at 25°C, time had no effect on the% RNV. The highest % RNV (56.45%) was achieved at the end of 1 day. Fermentation at 35°C was influenced by the time, with the highest% RNV (61.11%) obtained at the end of 7 days. Available lysine and methionine increased substantially ($P < 0.01$) over the control when sorghum was fermented at 25°C or 35°C for 4 days. Sorghum fermented at 25°C and at 35°C had methionine content of 25.68 and 26.79 mg/g N, respectively, whereas the control contained only 11.25 mg/g N. Lysine level were 33.2 and 34.5 mg/g N in sorghum fermented at 25°C and 35°C, respectively, whereas the control contained 9.1 mg/g N. None of the available B-vitamins studied was altered significantly in the fermented samples (**Table 35**).

The grain sorghum is mainly consumed as *bhakari* and it is an age-old food habit of the sorghum consumers in India. Any improvement in the cultivar for making better quality bhakari will be appreciated. Considering the nutritional and evident quality problems of grain sorghum, the malting or fermentation of the whole sorghum flour for making bread, pancake, and other breakfast foods hold good promise of improving the functionality and utilization of sorghum as staple food by all classes of people in India. However, the utilization of sorghum in the form of beer or table wine seems to be uneconomical in comparison to the whole flour due to social and psychological food habits, low amylopectic power of sorghum malt and significant loss of solids during the beer making process.

According to Novellie (1968) the beer-making process may depend upon two fermentations involving (1) bacteria that produce lactic acid, (2) yeasts that produce alcohol. The malting process that generates the fermentable mono and disaccharides is depending upon the α- and β–amylases and the maltose splitting enzyme, maltase, which develops in the sorghum during germination. The beer is sour due to the presence of lactic acid (pH 3.3 to 3.5), opaque because of suspended solids and yeast, and lightly alcoholic due to 3 to 3.5 percent alcohol by mass. Sorghum beers differ in many ways from barley beer (**Table 36**). The nutritional value of sorghum beer is higher than that of most other beers because of the yeasts and other suspended material. The beer made from highly refined maize (adjunct) and unrefined sorghum malt has a lower vitamin value than that made from the same malt with ground whole sorghum grains as adjunct (**Table 37**). *Amgba* and *affouk* contained, respectively 20 to 31 per cent of the original sorghum solids. It was concluded that consumption of sorghum flour is more economical than that of sorghum beer or wine in terms of calories and proteins (**Table 38**).

Table 34: Composition of non-fermented and fermented sorghum meals*.

Constituent	Non-fermented	Fermented	Significance between meals
Proximate analysis (%)			
Crude protein	10.2	14.48	NS
Crude fiber	2.45	2.35	NS
Crude fat	3.13	3.20	NS
Ash	1.56	1.40	NS
Carbohydrates	82.84	82.57	NS
Available amino acids (mg/gN)			
Lysine	23.40	40.34	P < 0.001
Leucine	27.50	59.38	P < 0.001
Isoleucine	38.67	81.62	P < 0.001
Methionine	15.98	33.35	P < 0.001
% RNV	70.00	83.72	P < 0.001
Protein digestibility (%)			
With papain	1.42	8.33	P < 0.005
Without papain	1.36	8.21	P < 0.005

Constituent	Non-fermented	Fermented	Significance between meals
Carbohydrate availability (%)			
Reducing sugars	0.19	0.42	P < 0.001
Starch (with takadiastase)	16.70	31.82	P < 0.001
Starch (without takadiastase)	1.36	8.21	P < 0.001
Vitamins (mg/g)			
Niacin	37.90	41.30	P < 0.001
Thiamine	20.20	47.10	P < 0.001
Riboflavin	0.90	1.30	P < 0.005

*Means of four replications. Source: Kazanas and Fields, 1981.

Table 35: Nutritional processing quality of sorghum.

Nutrients	Sorghum meal		
		Fermented	
	Control	25°C	35°C
Methionine (mg/gN)	9.1a*	33.2b	34.5b
Lysine (mg/gN)	11.25 a*	25.68b	26.79b
Thiamin (mg/gN)	3.66ab*	3.18b	3.87b
Riboflavin (mg/gN)	1.34a*	1.27a	1.38a
Niacin (mg/gN)	68.39a**	70.88a	70.91a
% RNV	45.57**	55.10a	56.17a

N = 5, Means with different letters are significantly different.
*Significant at P<0.01; **Significant at P < 0.05; Source: Au and Field, 1981.

Table 36: Composition of sorghum beer with that of European (Barley Beer).

Beer	Raw materials used			Fermentation process	
	Malt	Adjunct	Hops	Lactic	Alcoholic
Sorghum	Sorghum malt	Maize or sorghum	None	+	+
European	Barley malt	Cereals or syrups	+	-	+

Source: Novellie, 1977.

Table 37: Vitamin content (µg/100 g) of sorghum beer.

Vitamin	1975 Maize as adjunct	1976 Sorghum as adjunct	1972	1973
Thiamine	27	47	35	32
Niacin	222	400	270	189
Riboflavin	27	35	27	36

Source: Novellie, 1977.

Table 38: Comparison of nutritional composition of sifted flour, *amgba* and *affouk* as% of whole grain.

Composition	Sorghum	Sifted flour	Amgba	*Affouk* wine
Solids	100	70	20	31
Calories (Alcohol not included)	100	75	20	33
Total calories (Alcohol included)	100	75	29	55
Protein	100	69	19	41
Lysine	100	-	41	56
Lipid	100	35	0	28
Total carbohydrates	100	72	19	29
Cellulose	100	28	0	12
Ash	100	39	36	42
Calcium	100	59	35	81
Total P	100	41	38	65
Phytic P	100	-	13	47
Thiamine	100	39	160	65
Riboflavin	100	34	145	141
Niacin	100	30	35	41

Source: Chevassus-Agnes *et al.* (1976)

Effect of fermentation on soluble proteins and *in-vitro* protein digestibility of sorghum, green gram and sorghum + green gram blends

Sorghum seeds (cv. M 35 –1) were obtained from the Senior Sorghum Breeder, MPKV, Rahuri and the green gram (*Vigna radiata*) seeds were purchased from the local market. Grains were separately milled to 60 mesh and mixed (70:30, w/w) to obtain composite meal. The meals were mixed with water (1:3, w/v) and natural fermentation was carried out up to 5 days at 25°C in beakers covered with aluminium foil. Sorbic acid (0.1% in meal) was used to prevent mold growth during fermentation. The pH of the fermenting material was measured with Elico-pH meter while the titratable acidity expressed as percent lactic acid, was determined by titration with 0.1N NaOH. The slurries were dried in petri dishes at 55°C to constant weight and powdered. The dried meals were analyzed for crude protein, soluble proteins, free amino acids and in-vitro protein digestibility.

Fermentation at 25°C upto 5 days were monitored and changes in pH, titratable acidity and proteins of meals of sorghum, green gram and sorghum + green gram blend (70:30, w/w) were studied. The pH of fermenting meals decreased sharply with concomitant increase in the titratable acidity within 24 hrs. The proteins, free amino acids, soluble proteins and in-vitro protein digestibility markedly increased within 24 hrs fermentation (**Table 39** and **40**). Supplementation of green gram with sorghum decreased the acidity and improved the contents of crude protein, soluble proteins and free amino acids of fermented sorghum meal.

Table 39: Effect of fermentation on pH, acidity and protein of sorghum, green gram and sorghum-green gram blend.

Fermentation period (days)	PH			Titratable acidity (%)			Protein (N x 6.25) (%)		
	Sorghum	Green gram	Sorghum + green gram	Sorghum	Green gram	Sorghum + green gram	Sorghum	Green gram	Sorghum + green gram
Control	6.3	5.9	6.1	0.11	0.33	0.14	11.4	24.6	14.9
1	4.3	4.5	4.1	0.44	0.43	0.46	12.8	26.8	17.0
2	3.7	4.3	3.9	1.01	0.44	0.62	13.1	27.0	17.2
3	3.6	4.3	3.8	1.15	0.45	0.79	13.2	27.2	17.6
4	3.5	4.3	3.7	1.27	0.46	0.86	13.5	26.6	17.1
5	3.5	4.2	3.7	1.36	0.47	1.00	13.6	26.5	16.8

Effect of malting and fermentation on nutritional quality of sorghum

The grains of sorghum cultivars M 35 –1 and Swati and CSH – 5 were steeped in water for 10 hrs, germinated at 30°C for 24 hrs in incubator, dried at 60°C to constant weight and milled to 60 mesh. For fermentation studies the grains were milled to 60 mesh, mixed with distilled water (1:3, w/v) containing 0.1% sorbic acid to prevent mold growth and fermented in beakers covered with aluminium foil for 24 hrs at 25°C. The slurries were dried to constant weight at 60°C. The dried meals were analyzed for crude protein, free amino acids, water soluble proteins, starch, in-vitro protein digestibility and loss in dry weight was measured by weighing the dry meals before and after malting or fermentation.

Table 40: Effect of fermentation on soluble proteins, free amino acids and in-vitro protein digestibility of sorghum, green gram and sorghum-green gram blend.

Fermentation period (days)	Soluble proteins (%)			Free amino acids (mg/g meal)			IVPD (%)		
	Sorghum	Green gram	Sorghum + green gram	Sorghum	Green gram	Sorghum + green gram	Sorghum	Green gram	Sorghum + green gram
Control	1.4	11.7	5.1	2.6	4.8	2.7	80.2	82.0	80.8
1	2.1	15.3	6.2	6.9	22.4	12.5	85.8	87.0	86.2
2	2.1	15.3	6.4	7.1	23.2	14.8	85.4	87.1	86.3
3	2.1	16.2	6.3	7.5	23.5	14.9	87.4	87.2	86.4
4	2.1	15.8	6.5	7.9	24.0	15.8	86.9	87.9	87.3
5	2.1	15.5	6.1	8.2	25.1	17.8	85.7	88.3	86.4

Both malting and fermentation treatments markedly improved the nutritional quality of sorghum (**Table 41**). The reducing sugars, free amino acids and protein digestibility were found to be increased after malting of grains for 24 hrs. Malting however, did not affect the contents of crude or soluble proteins. The fermentation of meals for 24 hrs markedly increased both the crude and soluble proteins along with reducing sugars, free amino acids and protein digestibility. Natural fermentation of meal appears to be superior processing technique over malting in improving the nutritional quality of grain sorghum. The loss in dry matter both during malting and fermentation was higher in CSH – 5 than Swati and M 35 – 1.

Table 41: Effect of malting and fermentation on nutritional quality of sorghum cultivars.

Parameter/Cultivars/ Malting/Fermentation period	M 35 - 1		Swati		CSH - 5	
A. Malting	0	24	0	24	0	24
Malting loss, %	-	3.2	-	2.9	-	3.8
Crude protein, %	11.6	11.6	10.7	10.7	11.3	11.3
Soluble proteins, %	1.2	1.0	0.9	0.9	1.0	0.9

Parameter/Cultivars/ Malting/Fermentation period	M 35 - 1		Swati		CSH - 5	
Free amino acids, mg/g	2.2	3.3	1.7	3.3	1.8	2.6
Starch, %	69.5	68.0	68.2	67.0	68.4	66.3
Reducing sugars, %	0.2	1.4	0.1	0.7	0.1	1.4
IVPD, %	80.8	82.7	81.6	84.7	78.5	79.2
B. Fermentation						
Dry matter loss, %	-	1.6	-	4.4	-	6.6
Crude protein, %	11.4	12.8	11.7	13.2	10.2	11.8
Free amino acids, mg/g	2.6	6.9	1.7	6.2	1.7	6.3
Soluble proteins, %	1.4	2.1	1.4	2.1	1.1	2.0
Starch, %	76.3	74.0	72.1	69.5	74.4	73.5
Reducing sugars, %	0.3	4.1	0.2	3.1	0.2	3.2
IVPD, %	80.2	85.8	79.4	87.4	78.6	84.6

Utilization of malted sorghum, mung bean and black gram in preparation of *bhakari / roti*:

The grains of sorghum (CSH – 5), mung bean and black gram were malted for 24 hrs at 27°C, dried at 60°C and milled to 60 mesh. The meals of malted grains were separately blended with normal sorghum flour at different proportions. The *bhakari* were prepared from 50g of such composite meal. The fresh *bhakaries* were evaluated for their sensory properties by the semi-trained judges using 10-point hedonic scale. The average score values given by 5 judges were used to determine acceptability of the product.

Blending of the meal of malted sorghum with normal flour to the extent of 80% lowered the flavour and enhanced the sweetness of *bhakari*. Blending of malted mung bean flour to the extent of 30% with sorghum produced acceptable *bhakari* with significant improvement in sweetness as compared with unmalted mung bean:Sorghum blend. Similar results were also obtained with unmalted and malted black gram (**Table 42**). An improvement in sweetness of the product can be attributed to the increased sugar levels in malted meals. The results indicate that malted sorghum to the extent of 80% and mung bean or black gram to the level of 30% can be blended with normal sorghum flour to obtain acceptable *bhakari* with improved nutritional value.

Table 42: Sensory qualities of *bhakari* prepared from sorghum meal supplemented with malted sorghum, mung bean and black gram flour.

Flour blend (w/w)	Colour and appearance	Flavour	Texture	Taste/Sweetness	Overall mean
Sorghum:Malted sorghum					
100:0	7.0	7.3	7.7	7.5	7.4
80:20	6.5	6.2	7.0	7.7	6.9
60:40	7.0	6.5	7.2	7.7	7.1
40:60	6.5	6.3	7.3	7.7	7.0
20:80	6.8	6.3	7.5	8.0	7.2
S. E. ±	N.S.	N.S.	N.S.	N.S.	N.S.
Sorghum:Normal mung bean					
100:0	8.3	8.1	8.3	8.5	8.3
90:10	8.3	8.1	7.8	7.5	7.9
80:20	7.9	8.2	7.4	7.4	7.7
70:30	7.4	7.6	7.3	7.1	7.4
CD at 5%	N.S.	N.S.	N.S.	N.S.	N.S.
Sorghum:Malted mung bean					
100:0	8.0	8.0	8.0	8.1	8.0
90:10	8.0	7.8	7.8	8.1	7.9
80:20	7.8	7.4	7.6	8.1	7.7
70:30	7.8	7.8	7.6	8.1	7.8
S. E. ±	N.S.	N.S.	N.S.	N.S.	N.S.
Sorghum:Normal black gram					
100:0	8.5	8.2	8.3	8.5	8.4
90:10	8.3	8.0	7.5	7.8	7.9
80:20	7.8	8.0	7.5	7.3	7.7
70:30	7.5	7.5	7.3	6.5	7.2
CD at 5%	N.S.	N.S.	N.S.	N.S.	N.S.
Sorghum:Malted black gram					
100:0	7.6	7.6	7.4	7.4	7.5
90:10	7.6	7.8	7.8	7.6	7.7
80:20	7.8	8.0	8.0	8.2	8.0
70:30	8.0	7.6	8.0	8.4	8.0
S.E. ±	N.S.	N.S.	N.S.	N.S.	N.S.

N.S. = Non-significant

Effect of malting on proximate composition and *in-vitro* protein digestibility and starch digestibility of grain sorghum:

The changes in dry matter, nutrient composition and *in-vitro* protein and starch digestibilities were studied during malting of grain sorghum. The steeping of grains in water for 10 to 30 hrs and germinating at 30ºC upto 72 hrs significantly decreased the dry weight and starch while significantly increasing the water soluble proteins, free amino acids and reducing sugars in malted meal (**Table 43** and **44**). The steeping of grains for 10 hrs and subsequent germination for 24 hrs significantly improved *in-vitro* protein and starch digestibility of the malted meal. However, prolonged steeping or germination of grains exhibited adverse effects on the digestibility of starch. The results indicated that 10 hrs steeping followed by 24 hrs germination is optimum to produce sorghum malt for traditional food uses.

Table 43: Effect of malting on malting loss (%) and proximate composition of sorghum.

Composition	Steeping period (hr)	Malting period (hr)		
		24	48	72
Malting loss (%)	10	3.2	11.2	19.5
	20	6.2	13.6	20.9
	30	9.5	17.5	25.6
Crude protein (%) (11.6)	10	11.6	11.3	11.2
	20	10.6	10.5	10.1
	30	10.2	10.2	9.9
Crude fat (%) (4.0)	10	4.0	3.0	3.0
	20	4.1	3.2	2.8
	30	4.2	3.2	2.6
Crude fiber (%) (1.8)	10	2.0	1.8	1.7
	20	1.9	1.8	1.8
	30	1.8	1.7	1.6
Ash (%) (1.7)	10	1.0	1.0	1.0
	20	1.0	0.9	0.9
	30	1.2	1.0	1.0

LSD at 5%, Malting loss = 0.05; Protein = 0.3; fat, fiber and ash = 0.2.
Figures in parentheses represent the composition of unmalted dry grain.

Table 44: Effect of malting on soluble proteins, free amino acids, starch, reducing sugars and in-vitro protein and starch digestibilities in sorghum.

Constituent	Steeping period (hr)	Malting period (hr)		
		24	48	72
Soluble proteins (%) (1.2)	10	1.2	1.6	1.9
	20	1.3	1.9	2.4
	30	1.9	2.3	2.8
Free amino acids (mg/g) (2.2)	10	3.3	5.9	7.1
	20	3.5	6.5	8.1
	30	5.8	7.8	10.2
Starch (%) (69.5)	10	68.0	60.5	57.6
	20	64.3	55.1	53.1
	30	62.3	50.5	49.1
Reducing sugars (%) (0.2)	10	1.4	2.7	4.8
	20	1.7	3.7	6.4
	30	2.6	4.7	7.3
IVPD (%) (80.8)	10	82.7	85.3	90.2
	20	79.5	80.9	83.4
	30	78.4	79.4	83.8
IVSD (mg maltose/g/2hr) (137.1)	10	204.0	125.7	97.1
	20	136.4	124.7	76.0
	30	121.4	78.5	60.7

LSD at 5%, Soluble proteins = 0.2; Free amino acid = 3.2; starch = 2.8; Reducing sugars = 0.2; IVPD = 2.5; IVSD = 2.8.
Figures in parentheses represent the composition of unmalted dry grain.

SUPPLEMENTING SORGHUM FLOUR WITH WHEAT MAIDA

Effect of incorporation of sorghum flour in wheat maida on physical and sensory properties of bread, buns, cookies and biscuits.

Bread, sweet buns, cookies and orange biscuits were prepared from the composite flour of wheat and sorghum. The products were evaluated for their physical and sensory properties. The results revealed that sorghum flour with 80% extraction could be mixed with maida to the extent of 20% (w/w) to prepare acceptable bread and buns and to the extent of 30% (w/w) to prepare acceptable cookies and biscuits without using any other additives (**Table 45, 46** and **47**).

Table 45: Sensory properties of bread, sweet buns, cookies and orange biscuits prepared from wheat : sorghum blends.

Product/Blend	Flavour	Crust colour	Crumb or texture	Taste (Sweet-ness)	Overall acceptability	Spread factor (w/t)
Bread wheat:sorghum						
100:0	8.00	8.50	9.50	9.50	8.88	-
90:10	7.50	8.00	6.50	7.50	7.38	-
80:20	7.50	7.00	6.50	7.00	7.00	-
70:30	5.50	6.50	4.50	5.00	5.38	-
S. E. ±	0.24	0.41	0.36	0.28	0.32	-
CD at 5%	0.78	1.33	1.18	0.90	1.05	-
Sweet buns						
100:0	8.00	8.00	8.00	8.00	8.00	-
90:10	8.00	8.00	8.00	8.00	8.00	-
80:20	7.50	8.00	6.50	8.00	7.50	-
70:30	6.00	6.50	6.00	6.50	6.26	-
S. E. ±	0.13	0.13	0.13	0.13	0.13	-
CD at 5%	0.41	0.41	0.41	0.41	0.41	-
Cookies						
100:0	7.70	7.25	7.35	7.30	7.40	6.45
90:10	6.75	6.75	7.25	7.00	6.94	6.62
80:20	6.25	6.25	6.50	6.75	6.44	6.70
70:30	5.85	6.50	6.50	6.75	6.40	6.71
60:40	3.50	6.00	5.00	6.00	5.13	6.48
50:50	3.00	6.00	5.50	5.75	5.06	6.47
40:60	3.00	6.00	5.25	5.50	4.94	6.44
S. E. ±	0.44	0.30	0.35	0.35	0.36	0.013
CD at 5%	1.32	0.90	1.04	1.06	1.08	0.040
Orange biscuits						
100:0	7.50	7.75	7.25	7.50	7.50	8.40
90:10	6.75	7.50	7.25	7.50	7.25	8.61
80:20	6.25	6.75	6.50	7.50	6.75	8.70
70:30	6.20	6.75	6.25	7.50	6.68	8.80
S. E. ±	0.35	0.29	0.26	0.24	0.29	0.007
CD at 5%	1.15	0.94	0.86	0.77	0.93	0.023

Table 46: Physical properties of the bread and sweet buns prepared from wheat:sorghum blends.

Product/Blend	Volume (ml)	Vol./Wt. (ratio)	Weight of bread (g) during storage period			
			0 hr	24 hrs	48 hrs	72 hrs
Bread						
Wheat:Sorghum						
100:0	960	2.65	362.00	346.50 (4.3)	346.00 (4.4)	344.80 (4.8)
90:10	945	2.60	363.00	347.50 (4.3)	347.00 (4.4)	345.50 (4.8)
80:20	905	2.46	367.80	353.00 (4.0)	349.00 (5.1)	346.00 (5.9)
70:30	900	2.43	370.00	356.00 (3.8)	351.00 (5.1)	349.50 (6.5)
Sweet buns						
Wheat:Sorghum						
100:0	140.50	3.07	45.70	45.00 (1.5)	44.80 (2.0)	43.90 (3.9)
90:10	140.00	2.99	46.90	46.00 (1.9)	45.70 (2.6)	45.00 (4.1)
80:20	132.00	2.78	47.50	47.00 (1.1)	46.60 (1.9)	45.20 (4.8)
70:30	130.00	2.68	48.50	48.00 (1.0)	47.00 (3.1)	46.80 (3.5)

Figures in brackets indicate percent loss in weight during storage period.

Table 47: Protein and ash contents of the bakery products prepared from wheat: sorghum blends (dry weight basis).

Products	Protein (%)	Ash (%)
Bread		
Sorghum 100%	5.60	0.89
Wheat:Sorghum		
100:0	11.80	2.83
90:10	11.18	2.63
80:20	11.42	2.55
70:30	9.51	2.40

Products	Protein (%)	Ash (%)
Sweet buns		
Wheat:Sorghum		
100:0	9.13	1.80
90:10	9.10	1.75
80:20	8.00	1.72
70:30	7.90	1.71
Cookies		
Wheat:Sorghum		
100:0	6.19	2.09
90:10	5.33	1.74
80:20	4.95	1.64
70:30	4.95	1.54
60:40	4.57	1.40
50:50	4.19	1.33
40:60	4.10	1.30
Orange biscuits		
Wheat:Sorghum		
100:0	7.23	1.87
90:10	6.85	1.50
80:20	6.60	1.39
70:30	6.20	1.31

Convenience foods from sorghum

Alternative foods such as ready-to-cook mixes, infant and weaning foods, breakfast foods, snack foods, dehydrated products, flaked products, baked products and popped products can be prepared using sorghum flour and their semolina.

Infant food mixes/Weaning foods:

Ready-to-cook, *idli* and *dosa* mixes can be developed using sorghum flour. The details of the composite mixes available in the market and infant mixes tested in the research field are given in **Table 48** and **49**.

Table 48: Composition of *idli* and *dosa* mixes commercially available in the market.

Idli mix ingredients	Percent	*Dosa* mix ingredients	Percent
Dehulled sorghum (rava)	61 (40 mesh)	Sorghum flour	30 (60 mesh)
Black gram dal flour	30.5	Rice flour	30 (60 mesh)
Salt	3	Black gram dal flour	30
Sodium bicarbonate	2	Salt	3.3
Citric acid	2.7	Citric acid	3.85
Sodium acetate	0.65	Calcium carbonate	1.95

Infant and weaning mixes can be prepared using various types of legumes, which balance the nutritional requirement of the growing children. Some of the mixes/formulations, which were tested, are given below.

Table 49: Tested formulations of infant mixes.

Infant mix	Proportion of ingredients
Sorghum soya mix	
Sorghum rava : soybean flour : SMP	70:25:5
Sorghum soya mix	
Sorghum flour : soybean flour : sugar	70:10:20
Sorghum pigeonpea mix	
Sorghum flour : pigeonpea flour	80:20
Pearl millet pulse mix	
Pearl millet flour : green gram flour	70:30
Pearl millet flour : blackgram flour	70:30
Pearl millet : Bengal gram flour	70:30

Weaning foods are intended to supplement the diets of young children when breast milk is no longer sufficient in quantity and to serve as a bridge between breast milk and the daily diet of the young child. There are many commercial preparations available for feeding infants as weaning food. While formulating weaning foods, the nutrient requirements of the child should be taken into consideration. Children need proportionately more protein in relation to their calorie requirement because of their added protein needs for growth. Further, the infections, which are very common among underprivileged children of developing countries, increase their requirements for protein.

Attention must be given in weaning foods to the B-complex vitamins, particularly thiamine, riboflavin and niacin. Pyridoxine is usually adequate unless destroyed in processing. Folic acid, B_{12} may also sometimes need to be considered. Vitamin D is usually available in tropical countries because of the availability of sunlight. Minerals, iron and calcium must also be added. Iodine is essential for the proper functioning of the thyroid gland in the absence of which goiter develops. There must be some fat and oil in the diet to provide a small amount of linoleic acid.

For complete and proper utilization of the weaning foods, the nutrients should be present in the proportion required by the infants. An international agreement has been reached on the minimal compositional standards. The standards prescribed by the Codex Alimentarius Committee on Foods for Dietary Uses for complete infant food are presented in **Table 50**. It has been tentatively agreed that the maximum levels of nutrients shall be twice the minimum levels except for vitamin D. The vitamin levels are based on the recommendation of the Nutrition Committee of the National American Academy of Pediatrics, which, in turn, are based on levels found in human milk. It is, further, recommended that the proteins should be nutritionally equivalent to that of whole egg, i.e., the biological value should not be less than 70. In case of nutritionally poor proteins, the quantity is increased in proportion to their biological value. However, the maximum protein content should not exceed 3 g/100 K calories. The biological value of protein may be increased by addition of essential amino acids.

In India, the Central Food Technological Research Institute, Mysore, has developed a nutritious children's food called "Bal-Ahar" (**Table 51**). It is prepared by blending cereals (corn meal, wheat flour) and legumes (soy bean flour, groundnut flour, Bengal gram flour) in certain proportions with or without skim milk powder.

Table 50: Codex Alimentarius Proposed Standards 1971 (minimum amounts/100 available kilocalories).

Nutrient	Minimum requirement
Protein (g)	1.8
Fat (% of total energy content)	25
Linoleate (mg)	300
Vitamin A (μg)	75 (retinal), 450 (β-carotene)
Vitamin D (IU)	40
(μg)	1
Vitamin E (mg)	0.3
Vitamin B_1 (μg)	25
Vitamin B_2 (μg)	60
Vitamin B_6 (μg)	50

Nutrient	Minimum requirement
Pantothenic acid (μg)	300
Vitamin B_{12} (μg)	0.15
Calcium (mg)	50
Magnesium (mg)	6
Phosphorus (mg)	25
Iron (mg)	1
Iodine (μg)	5

Source: Bender, 1973.

Table 51: Nutritional specifications for Bal-Ahar.

Nutrient	Requirement
Protein	25 ± 4%
PER	> 2.0
Vitamin A	Not less than 1300 units/100g
Fat	2 to 6%
Fatty acids	Less than 0.02%
Vitamin B_1	Not less than 0.3 mg/100g
Vitamin B_2	Not less than 0.4 mg/100g
Niacin	Not less than 5 mg/100g
Total ash	Not less than 0.5%
Moisture	Maximum 10%
Crude fiber	Maximum 5%
Iron	Not less than 10 mg/100g
Calcium	Not less than 400 mg/100g

Source: Schrimshaw, 1974.

Snack and baked food

Deep fried foods are very popular in India in rural as well as urban sector. They have longer shelf life. Normally rice is the chief ingredient in the snack foods. Dehulled sorghum flour can be partly substituted in snack foods. Dehulled sorghum and maize can be flaked. In this process sorghum grains are pearled in a polishing machine. This pearled sorghum is wetted, steamed under pressure and then flaked in roller flakers. The flake thus prepared makes an excellent snack food.

Dehydrated foods from sorghum:

Usually *papads* are prepared with cereal and pulse combination. But *papads* can be prepared from dehulled sorghum flour (**Figure 14**).

Dehulled sorghum flour (25 g)

↓

Add chilli powder (7.5 g), salt (12.5g), cumin seeds (5g) and mix well

↓

Add lemon juice (1 tsp)

↓

Add the mixture to boiling water (500 ml) and cook for 10-15 min

↓

Remove from fire and knead well

↓

Make into small round balls

↓

Roll into *papads*

↓

Dry under sun or in dryer

↓

Papads

↓

Store in polythene bags

Figure 14: Processes for preparation of sorghum papads.

Selection of sorghum genotype for *papad* preparation

The grains of five *rabi* sorghum varieties *viz.,* RPASV-3, Pandhari Chickni, Dadar local, M-35-1, Phule Revati and two hybrids *viz.*, SPH-1620, CSH-15R were obtained from Senior Sorghum Breeder, All India Co-ordinated Sorghum Improvement Project, Mahatma Phule Krishi Vidyapeeth, Rahuri. Chemical analysis of sorghum grains for protein, total sugar, crude fiber, starch, amino acids, minerals, phenolics, ash and fat were done using NIR Spectrophotometer, Spectra Analyzer Serial No: 05; 281, ZEUTEC Opto Elektronik GmbH, Keiler 211, 24768 Rendsburg, Germany.

Cleaned sorghum grains were subjected to milling in laboratory by grinding mill. Whole sorghum flour was used for preparation of *papad*. Chemical analysis of *papad* for protein and fat were done using standard methods of AOAC (2000).

Standardization of formula (ingredients in g/kg) for preparation of sorghum papad: For standardization of formula various combinations were used and the best one selected for further utilization. Sorghum flour blended with ingredients salt, *papadkhar*, asafoetida, sesame seed, black pepper, water and made dough. This dough was utilized in the preparation of *papads*. All the standardized ingredients were kept constant throughout all treatments.

Ingredients	Formula No. 1	Formula No. 2	Formula No. 3	Formula No. 4	Formula No. 5
Salt	Suhana papad masala pack used (Market pack)	8	16	24	32
Papadkhar		2	5	8	10
Black pepper		1.20	6	4	1.20
Sesame seed		20	50	50	20
Asafoetida		0.30	1.4	4	0.50
Cumin seed		-	-	50	-
Water requirement (ml)		1000	1000	1000	1000

Preparation of dough and papads: Sorghum grain cleaned and ground in to flour then sieved to 60 mesh sieve and add ingredients (papdkhar + salt + sesame seed + asafoetida + cumin powder + black pepper; selected formula No. 5 as a standard by using organoleptic evaluation). Then mixed flour in boiling water, stirred well, mixed into smooth dough, divided dough into small portion (10-12g each ball). Then steamed the balls in steamer/pressure cooker at 15 psi for 10-15 min. Then placed these balls in plastic sheet, placed these balls in *papad* press machine and pressed into 0.10 or 0.15 cm thick sheet. After that *papads* dried at room temperature, packed in plastic bags and stored at room temperature for further study. Market sample *papads* of finger millet and black gram were used as a control.

Organoleptic evaluation of papad: Organoleptic evaluation of *papad* for colour and appearance, flavour, texture, taste and overall acceptability was carried out by using standard method of Amerine *et al.* (1965). For this 10 semi-trained judges were used and 1 to 9 point hedonic scale was used for rating the quality of the sorghum *papad*. The mean of ten judges was considered for evaluating the quality.

Physico-chemical characteristics of papad: Total number of *papads*, colour *of papads*, thickness of *papad*, diameter *of papad*, weight of *papad*, puffiness, crispiness, expansion of *papad*, bulk density parameters were measured using standard method of AOAC (2000).

Microbial quality and statistical analysis of papad: Total microbial count was recorded by standard plate count (SPC) method. The Glucose Yeast Extract

Peptone Agar was prepared and used as growth medium for yeast and potato dextrose agar medium was used for mould. The petridishes were incubated at 30 ± 2°C for 48 hr for counting fungal colonies. To take a colony count, magnifying lenses were used. The statistical design adopted to study the different aspects was Completely Randomized Design (CRD) using 7 treatments with 3-10 replications depending on the parameter studied. The results were tabulated and analyzed according to the procedure of Panse and Sukhatme (1965).

Nutritional constituents of sorghum grain: During the present investigation protocol developed for the preparation of *papad*. The most promising genotypes were screened to identify one suitable for the *papad* production. The nutritional composition of RPASV-3, Pandhari Chikni, Dadar local, M-35-1, Phule Revati, SPH-1620 and CSH-15R in respect of their various parameters are given here. The crude protein content in grain ranged from 10.28 to 11.37%. The variety M-35-1 recorded numerically higher level of protein (11.37%) in the grain and was at par with Phule Revati (11.36%), RPASV-3 (11.27%) and hybrid CSH-15R (10.77%; **Table 52**). The variation in the protein content among the sorghum genotypes is due to their genetic background/variability but it is non-significant. The results were in line with those obtained earlier workers (Beta *et al.* 1995; Chavan *et al.* 2009; Chavan and Patil, 2010). The fat content in grain ranged from 1.21-1.90%. The genotype Dadar local gave significantly higher level of fat (1.90%) in the grain which was at par with Pandhari Chikni (1.74%), M-35-1 (1.68%) and SPH-1620 (1.65%). The starch content in grain ranged from 58.56 to 61.56%. The genotype Pandhari Chikni gave numerically higher level of starch per cent (61.56%) in grain and at par with Dadar local (61.44%) followed by Maldandi (61.31%) and SPH-1620 (60.44%; **Table 52**). Chavan *et al.* (2009) reported starch content varied from 58.13 to 73.08 per cent. The carbohydrate in grain sorghum mainly constitutes the starch ranging from 56 to 75% (Subramanian and Jambunathan, 1984). The starch content is negatively and significantly correlated with protein content the physiochemical characteristics of starch are influenced by amylose content is significantly correlated with protein content, the physiochemical characteristics of starch are influenced by amylose content in sorghum (Miller and Burns, 1970). The starch content in hybrid ranged 74.41 to 75.95 per cent (Deshpande *et al.*, 2003). Present results are similar to the literature. The total sugar content in grain ranged from 1.12 to 2.47%. The variety M-35-1 gave significantly higher level of total sugar (2.47%) in the grain and at par with Dadar local (2.27%) followed by CSH-15R (1.78%) and SPH-1620 (1.76%). Ibrahim *et al.* (2010) recorded total soluble sugar content from 0.54 to 3.38%, from 0.54 to 4.89% and from 0.41 to 4.41% in Hamra, Shahla and Baida, sorghum varieties. The fibre content in grain ranged from 3.15 – 3.71%. The hybrid CSH-15R gave numerically higher level of fibre (3.71%) in the grain and at par with RPASV-3 (3.61%) followed by SPH-1620 (3.56%) and Dadar local (3.33%). Ratnavathi *et al.* (2000) reported crude fibre among the thirteen cultivars varied from 1.57% (M-35-1) to 2.4% (SPV-462). Vannalli *et al.* (2008) revealed that proximate composition of sorghum grain for crude fibre ranged from 1.21 to 2.74%. Chavan *et al.* (2009) reported crude fibre content ranged from 1.90 to 2.64%. The ash content in grain ranged from 2.33 to 3.07%. The hybrid CSH-15R gave significantly higher level of Ash (3.07%) in the grain and at par with SPH-1620 (3.01%) followed by Pandhari Chikni (2.84%) and Dadar Local (2.49%). All results for ash content are non-significant. This indicates that all genotypes are having similar amount of ash content which is good source for the mineral elements. Vannalli *et al.* (2008) revealed that proximate composition of sorghum grain for ash ranged from 1.14 to 1.72%. Yewale (2013) observed a range of ash content of sorghum grain ranged from 0.90 to 3.82 per cent. The phenolics content in grain ranged from 0.92 to 1.63%t. The genotype RPASV-3 gave numerically higher level of phenolics (1.63%) followed by Pandhari Chikni (1.54%) and Dadar local (1.34%). Now-a-days phenolics components which are coming under secondary metabolites act as a neutraceutical component. Therefore high phenolics content have high importance in the human nutrition. The content of phenolics in grain sorghum has been studied by several investigators. Variations in the phenolics content of grain sorghum from 4.8 to 8.2% (Harris *et al.*, 1970) and 2.69 to 6.88% (Jambunathan and Mertz, 1973) have been reported in high tannin cultivars. Mc Millian *et al.* (1972) observed a range of tannins from 0.12 to 0.8% in broad spectrum of available sorghum. The present results are in accordance with the literature.

Table 52: Nutritional composition of sorghum grains.

Genotype	Crude Protein (%)	Lipids (%)	Starch (%)	Total Sugars (%)	Fiber (%)	Ash (%)	Phenolics (%)
Varieties							
RPASV-3	11.27	1.63	58.96	1.12	3.61	2.36	1.63
Pandhari chikni	10.28	1.74	61.56	1.49	3.15	2.84	1.54
Dadar local	10.33	1.90	61.44	2.27	3.33	2.49	1.34
M-35-1	11.37	1.68	61.31	2.47	3.17	2.33	1.18
Phule Revati	11.36	1.21	58.56	1.46	3.36	2.47	1.24
Hybrids							
SPH-1620	10.43	1.65	60.44	1.76	3.56	3.01	0.92
CSH-15R	10.77	1.52	60.13	1.78	3.71	3.07	0.94
Range	10.28-11.37	1.21-1.90	58.56-61.56	1.12-2.47	3.15-3.71	2.33-3.07	0.92-1.63
Mean	10.83	1.61	60.34	1.76	3.40	2.65	1.26
SE ±	0.46	0.20	1.12	0.43	0.21	0.10	0.25
CD at 5%	N.S.	0.59	N.S.	1.31	N.S.	0.30	N.S.

N. S. = Non-Significant.

Amino acids profile of sorghum grain: The results of amino acids content in sorghum grains are represented in **Table 53**. Histidine content in the grain varies from 1.83 to 2.13 g/16 g N. The hybrid CSH-15R (2.13 g/16 g N) and variety M-35-1 (2.12 g/16 g N) recorded higher histidine content followed by Dadar local and Phule Revati (2.09 g/16 g N). The lowest histidine content was observed in RPASV-3 (1.83 g/16 g N). Histidine content reported by Moss *et al.* (1988), Robertson *et al.* (2006) and Chavan and Patil (2010) were 2.0 – 2.3, 1.2 - 2.2 and 2.06 g/16 g N, respectively. Isoleuoine content in sorghum grain ranged from 3.10-4.09 g/16 g N. The genotype Dadar local gave higher isoleucine content (4.09 g/16 g N) followed by M-35-1 (3.94 g/16 g N), hybrid SPH-1620 (3.10 g/16 g N) and Pandhari Chikni (3.36 g/16 g N) gave lowest content of isoleucine. Isoleucine content reported by Mosse *et al.* (1988), Robertson *et al.* (2006) and Chavan and Patil (2010) were 4.1-4.4, 3.2-4.7 and 3.83 g/16 g N, respectively. Leucine content in sorghum grain ranged from 11.67 to 13.75 g/16 g N. The genotype Dadar local gave higher level of leucine content (13.75 g/16 g N) followed by Pandhari Chikni and M-35-1 (13.46 g/16 g N). The genotype RPASV-3 recorded lower level of leucine content (11.67 g/16 g N). Leucine content by Mosse *et al.* (1988), Robertson *et al.* (2006) and Chavan and Patil (2010) were 12.7 to 14.7, 10.4 to 16.7 and 12.27 g/16 g N, respectively. Lysine content in sorghum grain ranged from 2.55-2.89 g/16 g N. The genotype Pandhari Chikni recorded higher level of lycine (2.89 g/16 g N) followed by M-35-1 (2.69 g/16 g N) and hybrid CSH-15R (2.66 g/16 g N). Hybrid SPH-1620 gave lower level of lycine content (2.55 g/16 g N). Lycine content reported by Mosse *et al.* (1988), Robertson *et al.* (2006) and Chavan and Patil (2010) were 2.1-2.5, 2.3-3.6 and 2.60 g/16 g N, respectively. Methionine content in sorghum grain ranged from 1.34 to 1.55 g/16 g N. The genotype Pandhari Chikni exhibited higher level of methionine content (1.55 g/16 g N) followed by RPASV-3 (1.43 g/16 g N) and M-35-1 (1.34 g/16 g N) gave lower level of methionine content. Methionine content reported by Mosse *et al.* (1988), Robertson *et al.* (2006) and Chavan and Patil (2010) were 5.2 - 5.7, 3.0 - 5.4 and 1.55 g/16 g N, respectively. Other all amino acids are also in the range of literature values.

Table 53: Amino acid content in sorghum grain (g/16 g N).

Genotype	Cys	Glu	Gly	His	Ile	Leu	Lys	Met	Phe
Variety									
RPASV-3	0.87	20.54	2.61	1.83	3.79	11.67	2.61	1.43	4.90
Pandhari chikni	1.14	20.82	3.24	1.98	3.36	13.46	2.89	1.55	4.59
Dadar local	0.70	21.49	2.31	2.09	4.09	13.75	2.61	1.36	5.02
M-35-1	0.72	21.32	2.29	2.12	3.94	13.46	2.69	1.34	5.00
Phule Revati	0.86	21.08	2.36	2.09	3.72	12.57	2.61	1.37	5.12
Hybrid									

Genotype	Cys	Glu	Gly	His	Ile	Leu	Lys	Met	Phe
SPH-1620	0.88	21.25	2.76	2.03	3.10	12.21	2.55	1.36	5.69
CSH-15R	0.87	21.37	2.48	2.13	3.76	13.28	2.66	1.35	5.07
Range	0.70-1.14	20.54-21.49	2.29-3.24	1.83-2.13	3.10-4.09	11.67-13.75	2.55-2.89	1.34-1.55	4.59-5.69
Mean	0.86	21.12	2.578	2.038	3.68	12.91	2.66	1.39	5.05
S.E. ±	0.13	0.31	0.312	0.09	0.313	0.716	0.102	0.06	0.30
CD at 5%	0.39	N.S.	N.S.	0.29	0.94	N.S.	0.30	0.20	0.91

Table 53 continued…..

Genotype	Pro	Ser	Thr	Trp	Tyr	Ala	Arg	Asp	Val
Variety									
RPASV-3	5.30	3.57	3.08	1.15	2.87	7.75	3.82	8.59	4.13
Pandhari chikni	4.96	4.12	3.30	0.76	3.44	5.48	5.69	9.04	3.33
Dadar local	6.12	3.14	2.96	1.05	2.77	7.84	5.21	6.69	5.15
M-35-1	6.12	3.05	2.92	1.12	2.68	8.03	5.23	6.65	5.35
Phule Revati	5.99	3.08	2.99	1.19	2.84	7.95	5.08	7.05	3.98
Hybrid									
SPH-1620	6.24	3.47	2.98	1.15	2.75	8.05	5.10	7.08	3.83
CSH-15R	6.18	3.17	2.95	1.08	2.87	8.02	5.40	6.56	4.21
Range	4.96-6.24	3.05-4.12	2.92-3.30	0.76-1.19	2.68-3.44	5.48-8.05	3.82-5.69	6.56-9.04	3.33-5.35
Mean	5.84	3.37	3.02	1.07	2.88	7.58	5.07	7.38	4.28
S.E. ±	0.46	0.35	0.12	0.13	0.23	0.86	0.54	0.93	0.66
CD at 5%	N.S.	N.S.	0.36	0.40	0.70	N.S.	1.64	N.S.	2.00

N.S. = Non-significant

Mineral content: Calcium content in the sorghum grain ranged from 21.77 to 36.44 mg/100g. The genotype Dadar local recorded higher level of calcium content (36.44 mg/100g) followed by Phule Revati (34.75 mg/100g) and M-35-1 (34.72 mg/100g. The genotype RPASV-3 exhibited lower level of calcium content (21.77 mg/100g). Anonymous (1995) reported calcium content in sorghum grain was 26 mg/100g. Chavan and Patil (2010) observed calcium content in sorghum grain was 25 mg/100g. Iron content in the sorghum grain ranged from 4.11 to 4.98 mg/100g (**Table 54**). The genotype RPASV-3 recorded higher level of iron content (4.98 mg/100g) followed by M-35-1 (4.88 mg/100g. Phosphorus content in different genotype of sorghum grain ranged from 491 to 517 mg/100g. The hybrid SPH-1620 had higher level of phosphorus content (517 mg/100g) followed by RPASV-3 (513 mg/100g). Potassium content in different genotype of sorghum ranged from 506 to 564 mg/100g. The genotype RPASV-3 recorded higher level of potassium content (564 mg/100g) followed by Phule Revati (556 mg/100g). Chavan and Patil (2010) observed potassium content in the sorghum grain was 537 mg/100g. Winchester and Makokha (2011) reported potassium content in the sorghum grain was 350-524 mg/100g. Zinc content in different genotype of sorghum grain ranged from 3.68 to 4.73 mg/100g. The genotype Pandhari Chikni showed higher level of zinc (4.73 mg/100g). Chavan and Patil (2010) observed zinc content in the sorghum grain was 3.91 mg/100g. Present results indicated that sorghum grains are good source of the mineral content which has very high importance in the human diet. Therefore inclusion of sorghum in daily diet will help to supply necessary amount of minerals to human body.

Table 54: Mineral composition of sorghum grains (mg/100 g).

Genotype	Ca	Fe	Mn	Mg	P	K	Cu	Na	Zn
Variety									
RPASV-3	21.77	4.98	1.77	238	513	564	0.90	23.63	4.12
Pandhari chikni	30.19	4.14	2.41	221	505	512	1.09	29.55	4.73
Dadar local	36.44	4.36	2.82	225	494	506	1.02	17.99	3.87
M-35-1	34.72	4.88	2.83	221	491	515	1.03	17.67	3.86
Phule Revati	34.75	4.15	2.76	223	498	556	0.90	25.54	3.75
Hybrid									
SPH-1620	34.50	4.74	3.04	224	517	535	0.93	24.38	3.68
CSH-15R	34.42	4.11	2.87	222	499	531	1.04	22.65	3.79
Range	21.77-36.44	4.11-4.98	1.77-3.04	221-238	491-517	506-564	0.90-1.09	17.67-29.55	3.68-4.73
Mean	32.39	4.43	2.64	224	502	531	0.98	23.05	3.97
S.E. ±	4.68	0.34	0.39	5.54	8.97	20.60	0.07	3.87	0.33
CD at 5%	14.05	N.S.	1.19	N.S.	N.S.	N.S.	N.S.	11.6	1.00

N.S. = Non-significant

Standardization of formula (ingredients g/kg) for sorghum papad: Standardization of formula for *papad* preparation used different ingredients in different proportion and made five formulae and prepared *papad* from these formulae. While formulating the ingredients various combinations were tried in the preliminary trials. During the formulations colour, texture, flavour, taste and overall acceptability of the sorghum *papad* were considered as the base parameters. For comparison following five formulas were taken and from these only one was selected as the best one (Formula No.5).

***Organoleptic evaluation of sorghum papad prepared from different five formulae*:** After organoleptic evaluation formula No. 5 was overall accepted in accordance with colour, texture, flavour and taste and it is selected for further *papad* preparation from all five varieties and two hybrids (**Table 55**).

Table 55: Organoleptic evaluation of various *papads* prepared by using five formulas.

Formula No.	Colour and appearance	Texture	Flavour	Taste	Overall acceptability	Ranking
Formula No.1	7.7	7.4	8.0	8.2	7.8	3
Formula No.2	8.2	8.4	8.3	8.5	8.3	2
Formula No.3	7.2	7.1	7.6	7.8	7.4	4
Formula No.4	6.6	6.7	7.0	7.2	6.8	5
Formula No.5	8.4	8.7	8.5	8.8	8.6	1

* 1 to 9 hedonic scale (Amerine *et al.*, 1965)

Physical characteristics of sorghum, black gram and finger millet papads: The total number of *papads* obtained from 1 kg of five varieties and two hybrids were weighed and numbers of papad are recorded from each genotype. The highest number of *papads* obtained from RPASV-3 (65), Pandhari Chikni (65) and Dadar local (65) genotype per kg and lowest number of *papads* obtained from hybrids SPH-1620 (57) and CSH-15R (56) per kg (**Table 56**). Colour of raw *papad* prepared from different five varieties and two hybrids recorded by visual observation. Among them the genotype RPASV-3 showed red brown coloured *papad*. The genotype Pandhari chikni recorded faint white coloured *papad*. The genotype Dadar local and M-35-1 gave faint yellow coloured *papad*. The variety Phule Revati and hybrid CSH-15R and SPH-1620 recorded yellowish coloured *papads*. Fingermillet *papad* gave faint yellow coloured and blackgram *papad* gave yellowish coloured *papad*. The colour of *papad* from each genotype varies according to their grain colour. This colour difference is from their genetic makeup. The genotype RPASV-3 exhibited faint red brown colour which is similar to that of finger millet. The market *papad* samples of finger millet are transport and it might be prepared from the finger millet extract (mostly starch) therefore, their puffing volume is more. The *papad* prepared from whole grain flour of RPASV-3 gave very good colour and appearance, so they can be used as a good snack food. The thickness of *papads* from all genotypes was measured by

Vernier Caliper in mm. Among that the papads prepared from Phule Revati (0.061mm) and CSH-15R (0.061mm) genotype recorded higher level of thickness followed by Pandahri Chikni (0.60mm) and RPASV-3 (0.059mm). *Papads* of finger millet gave 0.058mm and black gram 0.057mm thickness. The thickness of *papads* depends upon the preparation; because these all *papads* were prepared manually.

Table 56: Physical characteristics of sorghum, blackgram and finger millet *papads*.

Name of genotype	No. of papad/kg	Colour of papad	Thickness of papad (mm)	Diameter of papad (cm)		Expansion (%)	Weight of papad (g)		Increase weight (%)	Bulk density (in box) (g/cm^3)
				Before frying	After frying		Before frying	After frying		
RPASV-3	65	Red brown	0.059	16.5	34.54	22.2	16.2	19.1	17.90	0.045
Pandhari Chikani	62	Faint white	0.060	16.4	29.88	21.3	18.9	23.8	25.93	0.042
Dadar Local	65	Faint yellow	0.056	14.1	20.57	17.0	16.3	21.3	30.67	0.043
M-35-1	65	Faint yellow	0.057	15.8	27.21	20.1	19.4	23.4	20.62	0.043
Phule Revati	64	Yellowish	0.061	14.9	29.53	19.3	20.1	25.2	25.37	0.044
SPH-1620	57	Yellowish	0.058	16.4	10.97	18.2	20.0	25.2	26.00	0.043
CSH-15R	56	Yellowish	0.061	16.1	20.49	19.4	20.4	25.1	23.03	0.044
Finger millet (market sample)	-	Faint yellow	0.058	11.2	23.2	107.14	3.6	14.6	30.55	0.044
Black gram (market sample)	-	Yellowish	0.057	13.6	15.1	11.03	6.7	10.1	50.75	0.043

Diameter of papads: The diameter of raw *papad* ranged from 14.1 to 16.5cm and diameter of fried papad ranged from 10.97 to 34.54cm. The genotype RPASV-3 recorded significantly highest level of diameter both before (16.5cm) and after frying (34.54cm) followed by Pandhari Chickni (Before frying 16.4cm and after frying 29.88cm) and Phule Revati (Before frying 14.9cm and after frying 29.53cm. Diameter of *papad* is mostly dependent on the starch ability of the gelatinized starch quantity and quality present in the sample. Therefore, the raw and fried *papads* have different diameter. During frying of *papads* the starch present in the *papad* get expanded. The expansion of the starch is mostly dependent upon their structure and combination of amylose, amylopectin as well as sudden change during frying. Therefore these all *papads* gave different values for expansion depending upon above parameters. But this parameter gives the crispiness of the *papad*. Those *papad* have high expansion during frying gives more crispiness.

Weight of papad: The unit weight of raw and fried *papads* of different seven genotypes were recorded by taking the weight of ten *papads* of each variety and from that finally mean weight of *papad* was calculated. The mean weight of *papad* before frying ranged from 16.2 to 20.4g and mean weight of *papad* after frying ranged from 19.1 to 25.2g. The highest weight of *papads* before frying was observed in hybrid CSH-15 R (20.4g) followed by variety Phule Revati (20.1g). The highest weight of *papads* after frying was observed in variety Phule Revati (25.2g) and hybrid SPH-1620 (25.2g). The lowest weight was observed in genotypes RPASV-3 (19.1g) and Dadar local (21.3g). Weight of finger millet *papad* increased from 3.6 (before frying) to 14.6g (after frying) and in black gram *papad* weight increased from 6.7 (before frying) to 10.1g (after frying). Due to lack of research work on sorghum *papad* preparation as well as physical characteristics studies, the available literature found on sorghum flour preparation of with other millet flour for preparation of *papads* are given below. Nazni and Pradheepa (2010) prepared black gram *papads* with incorporation of sorghum at 0, 5, 10, 15 and 20% and other all ingredients were kept constant. They found that the samples were highly acceptable by the judges and physical properties were almost same with control *papads*. The present results showed that the whole sorghum flour can be used for the preparation of *papads* with good physical-chemical and nutritional quality properties.

Frying quality of *papad*

Oil absorption: Oil absorption capacity is an important characteristic to judge the frying quality of *papad*. Oil absorption capacity is determined by amount of oil absorbed by the *papad* after frying. For determination of oil absorption capacity weight of *papad* before frying and weight of *papad* after frying were taking into consideration. The oil absorption capacity was highest in genotype Dadar Local (30.67%) followed by SPH-1620 (26.00%). The oil absorption capacity was lowest in genotype RPASV-3 (17.90%; Table 5). In case of black gram (50.75%) and finger millet (30.55%) *papads* oil absorption capacity was highest than all seven genotypes of sorghum. While food product preparation higher oil absorption can create problem for that product during storage and reduce the shelf life. Therefore less oil required for *papad* frying is more beneficial. In case of sorghum *papads* of the genotype RPASV-3 required less amount of oil for frying and that gives additional benefit to the product. Now-a-days low oil/fat content in the diet is more beneficial for health. Therefore sorghum *papads* are far better than the other *papads* which are available in the market i.e. black gram and finger millet.

Puffiness: Puffiness is the frying quality which determines the expansion of *papad* after frying. Among the all seven genotype, the genotype RPASV-3 was highest Puffiness (22.2%) followed by Pandhari Chikni (21.3%) and M-35-1 (20.1%). The genotype Dadar Local showed lowest puffyness (17.00%) than all seven genotypes. As compare with finger millet *papads*, it gave highest puffiness (107.14%) but black gram *papads* gave lowest puffiness (11.03%).

Crispiness: Crispiness is the frying quality which shows the rheological characteristics of the product during sensory evaluation by mouth feel. Among the five genotypes and two hybrids the genotype RPASV-3 was scored highest crispiness to mouth feel. Finger millet *papad* was scored highest crispiness to mouth as compared with all seven genotypes of sorghum *papad* and black gram *papad*. Crispiness of the food product mostly based on the puffing of starch during frying. If the puffiness is higher than it gives very smooth and crunchy type crispiness. Crispiness of the snack food products is mostly acceptable by the all type of consumers. Among the sorghum genotype RPASV-3 gave good quality *papads* with higher crispiness than other genotypes. As compare with finger millet *papads* its crispiness is lower than finger millet *papads* due to less amount of puffiness by lower amount of starch content. Finger millet *papads* mostly prepared from the extract, so it contain higher amount of starch than sorghum. But while comparing all organoleptic parameters RPASV-3 genotype gave comparable *papad* quality to the finger millet.

Expansion: The mean expansion percentage of all seven genotype ranged from 17.00 to 22.2%. Among the sorghum *papads* the expansion% of genotype RPASV-3 was significantly higher (22.2%) followed by Pandhari Chikni (21.3%) and M-35-1 (20.1%). Finger millet papad gave higher expansion 107.14% among all sorghum and black gram *papads*.

Bulk density: Among the all seven genotype RPASV-3 had highest bulk density (0.045g/cm^3) followed by Phule Revati (0.044g/cm^3) and CSH-15R (0.044g/cm^3). As compare with finger millet *papads* it gave lowest bulk density (0.044g/cm^3) than genotype RPASV-3 (0.045g/cm^3). The black gram *papads* gave lowest bulk density (0.043 g/cm^3) than Phule Revati (0.044g/cm^3) and hybrid CSH-15R (0.044g/cm^3).

Organoleptic evaluation of sorghum papads: Colour is very important parameter in judging the property of fried *papad* that not only reflects the suitable raw material used for the preparation but also provides information about the formulation and quality of product. The mean colour and appearance sore of all seven genotype were different from each other. The score ranged between 6.4 to 7.4 with mean 7.0. The genotype Pandhari Chikni (7.4) and hybrid SPH-1620 (7.4) scored highest for colour and appearance followed by the genotype RPASV-3 (7.2; **Table 57**). Colour of any food material mostly important for the attraction of the consumer. The colour of *papad* prepared from the genotype RPASV-3 have distinct faint red brown colour, which was more attractive than other *papads* prepared during this study. This colour of *papad* is due to the reddish brown sorghum grains of RPASV-3. Consumer can easily identify the *papads* prepared from this sorghum variety in the market. Therefore this sorghum variety has this advantage for specific niche/snack product preparation.

Texture/Crispiness: Texture is combined sensation of all the rheological and structural parameters of the product during chewing and biting. It includes handfeel and monthfeel. The mean texture score of all seven genotype ranged from 6.4 to 7.8. The genotypes RPASV-3 scored highest (7.8) and Dadar local reported lowest score (6.4) among the genotypes.

Table 57: Organoleptic evaluation of sorghum *papad**

Genotype	Colour and appearance	Texture/ crispiness	Flavour	Taste	Overall accep-tability
Variety					
RPASV-3	7.2	7.8	7.0	7.8	7.6
Pandhari Chikni	7.4	7.2	7.6	7.6	7.4
Dadar local	7.0	6.4	7.0	6.8	6.6
M-35-1	6.6	6.6	6.8	6.2	6.4
Phule Revati	6.4	6.8	7.0	6.6	6.6
Hybrid					
SPH-1620	7.4	7.2	7.0	7.0	7.1
CSH-15R	7.0	7.2	6.8	7.2	7.0
Range	6.4-7.4	6.4-7.8	6.8-7.6	6.2-7.8	6.4-7.6
Mean	7.0	7.02	7.02	7.07	6.85
S.E. ±	0.35	0.43	0.25	0.52	0.44
CD at 5%	N.S.	N.S.	0.74	1.55	N.S.

N.S. = Non-significant*; 1 to 9 hedonic scale (Amerine *et al.* 1965)

Flavour: Flavour is main criteria that make the product to be liked or disliked. The perception of flavour is a combination of taste, smell, impression and texture. The flavour scored of all seven genotype ranged from 6.8 to 7.6 with mean 7.02. The genoypte Pandhari Chickni scored highest (7.6) followed by RPASV-3 (7.0), Dadar local (7.0), Phule Revati (7.0) and SPH-1020 (7.0). The lowest score was M-35-1 (6.8) and CSH-15 R (6.8). Flavour is also very sensitive parameter for acceptance for the food product. If the food product is giving pleasant flavour consumer accept that food product without any hesitation. Flavour is mostly depends on the frying oil quality for specially fried *papads* as well as some chemical reactions occur during frying. The new sorghum genotype RPASV-3 gave very pleasant flavour to the fried papad which was highly acceptable to the consumers (judged by the judges).

Taste: Taste of food product is sensation perceived by the taste buds and influenced by the texture, flavour, taste and composition of product. It is one of the essential parameter related to acceptability of the food product. The taste score for seven genotype of sorghum ranged from 6.2 to 7.8 with mean of 7.07. The significantly highest taste score was observed for genotype RPASV-3 (7.8) followed by Pandhari Chikni (7.6) and CSH-15R (7.2). The lowest score was in M-35-1 (6.2) which is on par with Phule Revati (6.6). Taste of food product is the prime parameter for consumers' acceptance. Those food products gave very pleasant, sweet and freshness enhancer taste are mostly preferred by the consumers. As a snake product sorghum papad prepared from RPASV-3 gave highest acceptability as compare to other papads prepared during this research study.

Overall acceptability: Overall acceptability is the total reflection of the scores obtained for colour, texture/crispiness, flavour and taste of the *papad*. The overall acceptability scored ranged between 6.4 to 7.6 with mean 6.85. The overall acceptability scores were highest for genotype RPASV-3 (7.6) followed by Pandhari Chikni (7.4) and CSH-15R (7.0). There is no research work done on sorghum *papad* preparation as well as their organoleptic properties studies. So no literature on this aspect is available. But other literature found on sorghum flour with other millet flour *papads* are given here. Nazni and Pradeepa (2010) the mean acceptability score obtained by the sensory evaluation of jawar *papads*. Among the different variation of jawar *papad*, control has got a highest score 8.80 followed by the variation S_2 with a score of 8.30 and the least score 7.90 is obtained by both the variation S_4 and S_5 for appearance attributes. Regarding the colour attributes the highest score 8.90 is obtained by control followed by variation S_2 with the score of 8.10. The texture attribute was found to be maximum for the control with a score of 8.80 and followed by the variation S_2 with the score of 7.50. Regarding the taste attribute, the highest score of 9.00 is obtained by the control which is followed by the variation S_4 with the scores of 8.10. The overall acceptability of control ranked first with a score of 8.90 and is followed by the variation S_4 with a score of 8.30.

Chemical composition of sorghum papad: The crude protein content in sorghum *papad* ranged from 10.11 to 11.35%. The *papad* prepared from M-35-1 gave numerically higher level of protein percentage followed by Phule Revati (11.25%) and RPASV-3 (11.14%; **Table 58**). The crude oil content in sorghum *papad* ranged from 23.59 to 35.42%. The hybrid SPH-1620 (35.4%) gave numerically higher level of oil percentage followed by M-35-1 (32.5%) and Phule Revati (32.2%). The genotype RPASV-3 (23.6%) recorded lower level of oil percentage. The protein content was highest in black gram *papad* (22.0%) followed by RPASV-3 (11.2%) *papad* genotype. The oil content was highest in finger millet *papad* (57.7%) followed by Black gram *papad* (34.4%) and lower oil content was in RPASV-3 variety (23.6%) of sorghum *papad*. The oil per cent in sorghum *papad* variety RPASV-3 was less as compared to black gram and finger millet *papad* hence the *papad* prepared from genotype RPASV-3 is good for consumption.

Comparison of sorghum papads with black gram and finger millet papads: Sensory evaluation of *papad* sorghum genotype RPASV-3 was done with finger millet and black gram *papads* (**Table 59**). Finger millet *papad* was overall accepted at rank 1 in accordance with colour and appearance, texture, flavour and taste followed by sorghum *papad* prepared from the genotype RPASV-3 and black gram *papad* but they are statically non-significant.

Table 58: Chemical composition of sorghum, finger millet and blackgram *papads*.

Name of genotype	Protein (%)	Oil (%)
RPASV-3	11.14	23.59
Pandhari chikni	10.20	30.48
Dadar Local	10.11	24.82
M-35-1	11.35	32.50
Phule Revati	11.25	32.18
SPH-1620	10.30	35.42
CSH-15R	10.33	25.42
Range	10.11-11.35	23.59-35.42
Mean	11.88	29.20
Finger millet *papad* (market sample)	9.10	57.72
Black gram *papad* (market sample)	22.00	34.35
S.E. ±	3.64	8.55
CD at 5%	N.S.	N.S.

N.S. = Non-significant

Table 59: Comparison of sorghum *papad* with black gram and finger millet *papad**

Name of genotype	Colour and appearance	Texture	Flavour	Taste	Overall acceptability	Ranking
Sorghum papad (RPASV-3)	8.5	8.2	8.0	8.2	8.23	2
Finger millet papad	8.8	8.4	8.3	8.5	8.50	1
Black gram papad	8.0	7.8	7.6	7.8	7.80	3

* * Semi-trained 10 judges used and 1 to 9 hedonic scales (Amerine *et al.*, 1965)

Microbial count of papad: Microbial studies were carried out after 3 month storage period of papads at room temperature. After taking observations, it was found that there was no any growth of yeast and mould.

Conclusion for sorghum papad

Mostly the local genotypes of sorghum *viz.,* Pandhari Chikni and Dadar local are used for preparation of *papads* in the cottage industry but at very small scale. The new genotype RPASV-3 identified for the preparation of sorghum *papads* showed slightly higher level of protein, fibre and pehnolics content. The total sugar, fat and ash content are less than other cultivars studied. *Papads* prepared from the genotype RPASV-3 had good sensory qualities like texture/crispiness, taste and overall acceptability and also good frying quality like puffiness, crispiness and expansion as compared to other genotype. *Papads* of genotype RPASV-3 having low oil absorption capacity (17.9%) than other six genotype and especially from market *papads* (black gram and finger millet *papads*). Due to less moisture percentage (2-3%) in sorghum *papad* there was no any microbial contamination. The product remains safe throughout the storage period. From the present study it is concluded that the genotype RPASV-3 is best suited for the *papad* snack product preparation.

Sorgos for syrup and sugar

The juicy, sweet-steamed sorgos have been used in sorghum growing countries for syrup and sugar production on laboratory scale. In this category 100 kg of stalks would yield 50 – 60 kg of juice. While preparation of sugar syrup the boiling should be stopped at 104°C. Boiling stop at the temperature of 108°C was reached. At this stage it would have a °Brix reading of 76 – 78°.

Sorghum grain for livestock feed

For ruminants: Digestibility of sorghum in cattle 74 while for barley is 84, in case of sheep it was 88% for sorghum and 77% for barley.

Poultry and pigs: The competitive price of mass-produced sorghum has made this an attractive feed grain for pig or poultry producers in the South-Western USA.

SPECIFIC PRODUCTS FROM SORGHUM

Sorghum *Hurda* (Sorghum Roasted Grains)

Procedure for the preparation of sorghum *hurda*

First select sorghum earheads of partially mature (soft/milk dough stage) tender stage. Roast the earheads directly on hot fire coke thoroughly. Separate the roasted grains by rubbing on the hard surface when hot. Add salt, lemon juice and chilli powder to taste. Serve it as such when it is hot. It can be preserved in LDPE bag by vacuum packaging for short time. The nutritive value of sorghum *hurda* is given in Table 47.

Sorghum cooked Grains (*Ghugarya*)

Ingredient	Quantity
Sorghum grains	250 g
Water	1000 ml
Salt	To taste

Procedure for the preparation of sorghum cooked grains

Take sorghum grains, clean properly and wash it thoroughly. Steep it in hot water for 24 hrs. After that drain it properly. Steam sorghum grains in pressure cooker under 15 Psi for 20 – 25 min. Then remove it from pressure cooker, mix with salt, spices as per requirement and then serve as hot or with any type of curries. The nutritive value of cooked sorghum grains is given in **Table 60**.

Sorghum *Khichadi*

Ingredients	Quantity
Pearled sorghum grains	100 g
Rice	25 g
Chickpea dhal	25 g
Groundnuts	50 g
Onion	2
Garlic	6 cloves
Fat	50 g
Mixed spices (mustard, cumin, asafetida, turmeric)	2 tea spoons
Salt	To taste
Water	As required

Procedure for the preparation of sorghum *Khichadi*

First heat the oil in a pot. Add spices and fry onion and garlic. Then add water and boil. When water starts boiling add pearled sorghum grains, rice, soaked chickpea, groundnuts and salt. Cover the pot and cook until grains become soft. Serve hot, garnished with grated coconut and green coriander leaves.

Sorghum *Pops*

Procedure for the preparation of sorghum *pops*

Pops can be prepared from sorghum grains, but it required special type of sorghum grains (special variety of sorghum), which gives very attractiveness, and fully open *pops*. First clean the sorghum grains. Wash it thoroughly. Sprinkle the grains with 2% hot water so as to attain 14 – 16% moisture and cover it for 4 – 6 hrs. Place iron skillet on gas and place salt in the skillet. After sufficient heating place the soaked sorghum grains and allow it to puff for about 5 min. (maximum pop grains). If the popping machine is available the soaked grains can be popped in the machine by adjusting their temperature and moving road distance in the machine. Remove the popped grains by sieving through perforated scoop by separating from salt. As per the requirement mix spices with the popped grains. Serve it when it is hot, it tastes crunchy and crispy. It can be packed immediately into HDPE pouch by vacuum packaging and store for about 6 months. The nutritive value of sorghum *pops* is given in **Table 60.**

Table 60: Nutrients available through various types products prepared from sorghum.

Product	Proteins (%)	Carbo-hydrates (%)	Fat/Oil (%)	Fiber (%)	Energy (Kcal)	Minerals (%)	Calcium (mg,%)	Iron (mg, %)	Vit. A (µg)
Sorghum flour low fiber	10.50	65.3	3.0	3.2	340	1.2	-	-	-
Sorghum flour high fiber	9.80	66.8	2.8	4.9	320	1.8	-	-	-
Sorghum *pop*	9.50	68.5	2.6	1.8	340	1.6	-	-	-
Sorghum *rava Semolina*	10.40	63.8	3.2	1.6	370	1.6	-	-	-
Sorghum *noodles*	11.3	56.7	1.6	0.8	335	1.2	-	-	-
Sorghum *noodles Kheer*	8.9	23.5	13.5	0.6	260	0.9	-	-	-
Sorghum *Modak*	12.5	60.5	1.8	0.8	330	1.1	-	-	-
Boiled sorghum	9.50	69.3	1.6	1.6	340	1.5	-	-	-
Roasted sorghum (*Hurda*)	8.00	70.2	1.5	1.4	350	1.3	-	-	-
Sorghum flakes	10.40	68.3	1.8	1.6	370	1.2	-	-	-
Sorghum flour for diabetic patients	15.60	61.3	1.9	6.4	435	2.8	-	-	-
Sorghum malt flour	14.50	69.8	2.3	1.5	343	0.8	-	-	-
Sorghum *roti / Bhakari*	12.64	66.8	2.1	1.8	343	1.8	18	9	-
Sorghum *dashami*	10.30	65.9	2.0	1.4	289	1.4	75	-	-

Product	Proteins (%)	Carbo-hydrates (%)	Fat/Oil (%)	Fiber (%)	Energy (Kcal)	Minerals (%)	Calcium (mg,%)	Iron (mg, %)	Vit. A (μg)
Malted sorghum flour *thalipeeth*	12.70	69.8	5.6	1.8	394	0.9	108	5	1237
Sorghum *upama*	9.40	56.8	5.7	1.6	326	1.1	61	4	-
Sorghum *Pakoda*	7.5	25.7	53.8	1.3	650	1.2	-	-	-
Sorghum *shira* (sweet)	8.73	58.3	15.6	1.7	432	0.8	32	6	-
Malted sorghum flour *papad*	13.00	69.8	1.6	2.8	354	3.4	43	58	-
High fiber *papad*	12.00	71.3	1.5	3.8	348	2.3	122	60	-
Malted mix *papad*	10.23	69.9	1.9	1.8	345	2.4	203	-	-
Sorghum-chilli papad (Spicy-papad)	10.00	69.5	1.6	1.8	361	1.6	198	-	-
Sorghum *idli*	8.00	65.4	1.8	1.8	200	1.8	103	-	-
Sorghum *dosa*	10.00	63.4	1.9	1.6	593	1.4	31	-	-
Sorghum *uttapa*	8.00	58.6	6.5	1.6	374	1.3	36	1.2	222, IU
Sorghum *appe*	10.00	55.8	3.4	1.5	248	1.8	105	3.0	-
Sorghum *khakara*	10.00	60.3	1.9	1.5	441	1.8	21	3.3	-
Sorghum *chiwada*	9.33	68.7	1.8	1.5	399	1.3	65	2.3	-
Sorghum-soy *ladoo*	17.00	65.8	2.5	1.5	648	2.2	98	3.2	-
Sorghum *chakali*	11.20	68.2	18.2	1.4	570	2.0	101	18.0	-
Sorghum *Kurdai*	10.40	40.1	21.5	1.4	349	1.2	25	1.3	-
Sorghum *papadi*	11.00	43.8	23.6	1.5	353	1.2	36	1.6	-
Sorghum masala *vada*	11.00	65.3	20.3	1.5	358	1.6	52	2.3	-
Sorghum *ambil*	13.50	57.6	1.8	1.4	409	1.4	174	65	-
Sorghum cookies	6.00	60.8	5.8	1.4	412	1.1	11	4	-
Malted sorghum cookies	6.00	68.3	1.9	1.5	421	1.1	13.3	3.2	-
Sorghum chocolate cookies	5.20	66.5	9.5	1.6	415	1.2	12.4	2.2	-
Cookies for diabetic patients	12.60	69.8	18.5	6.4	435	1.1	-	-	-
Sorghum *nankhatai* cookies	6.00	58.6	10.0	1.5	370	1.0	25	3.0	-
Sorghum milting moments	7.00	56.9	11.8	1.6	493	1.4	24	-	-
Sorghum *shankarpale*	6.50	58.6	22.3	1.4	381	1.4	17	2	-
Sorghum-groundnut cookies	7.50	42.8	15.3	1.5	464	1.4	21	2.5	565

Product	Proteins (%)	Carbo-hydrates (%)	Fat/Oil (%)	Fiber (%)	Energy (Kcal)	Minerals (%)	Calcium (mg,%)	Iron (mg, %)	Vit. A (μg)
Burffi	8.9	65.3	11.5	0.6	420	1.3	-	-	-
Sweet roles	5.6	38.9	3.5	0.8	230	1.2	-	-	-
Muffins	15.3	56.3	8.5	1.3	650	1.4	-	-	-
Salt biscuits	8.7	41.5	23.3	1.1	422	1.0	-	-	-
Sweet biscuits	7.5	57.1	27.3	0.9	510	0.6	-	-	-
Coconut biscuits	5.7	40.5	30.3	0.8	430	0.9	-	-	-
Peanut biscuits	7.5	65.5	9.3	1.5	465	1.4	21	2.5	560
Sorghum masala biscuits	6.20	63.3	2.9	1.4	407	1.4	33	3	-
Sorghum cake	8.00	60.2	12.6	1.5	418	1.4	88	-	-
Fruits cake	9.6	63.5	11.4	1.4	415	1.2	-	-	-

Two genotypes of *rabi* sorghum which are popular at local/village level for pops and one new genotype developed for specifically for sorghum *pops* (RSGV 3; Phule Panchami) and M 35-1 were grown at Sorghum Improvement Project, Mahatma Phule Krishi Vidyapeeth, Rahuri and maize grains were procured from the local market

Pops preparation

Fully mature grains of four sorghum genotypes and maize grains were used for the preparation of *pops*. Small popping machine, big popping machine (popcorn machine), micro-wave and gas fire-*bhatti* were used for the preparation of sorghum *pops*. Sample grains were tempered to 14% moisture content on a dry weight basis. The popping percentage was calculated on the basis of number of grains popped as well as on weight basis. The popped sorghum was quickly transferred to a measuring cylinder and the expansion at the highest point was recorded.

Nutritional and chemical analysis

The crude protein, hectoliter weight of grains, *pops* and expansion volume of sorghum *pops* was measured by using standard method of AOAC (1990). The colour of grains and size of pops was measured by physical observations. Starch content was determined using standard method of McCready *et al.* (1956). Total soluble sugars were determined using Dubois *et al.* (1956) method. Free amino acids were estimated by using standard method of Moor and Stein (1948). Amino acid profile, mineral content, vitamins and antinutritional factors in the sorghum grain, *pops* and pops flour were determined using NIR Spectrometer, SpectraAlyzer, Zeutec Make, Germany. Soluble proteins were determined using method of Lowery *et al.* (1951).

Sensory evaluation

The sorghum *pops* prepared were subjected for the sensory evaluation to a panel of ten semi-trained judges. Parameters evaluated by the judges included colour and appearance, texture, aroma/flavour, taste and overall acceptability. For the sensory evaluation of sorghum *pops*, 1 to 9 point hedonic scale [Like extremely (Excellent) – 9, Like very much (Very good) – 8, Like moderately (Good) – 7, Like slightly – 6, Neither likes nor dislike – 5, Dislike slightly – 4, Dislike moderately – 3, Dislike very much – 2, Dislike extremely – 1] was used (Amerine *et al.*, 1980). The NDF, ADF and IVDMD were determined from sorghum fodder by standard method of Van Soest (1963) and Van Soest *et al.*, (1967) respectively.

Statistical analysis

All analysis was conducted in triplicate and the average and their statistical analysis is given in the results as per the standard procedure of Panse and Sukatame (1967).

Fully mature grains of four sorghum genotypes and maize grains were used for the preparation of *pops*. Small popping machine, big popping machine (*pop*corn machine), micro-wave and salt-*bhatti* were used for the preparation of sorghum *pops*. In salt-*bhatti* method iron pan was kept on the gas and salt was added into it and temperature rose to 290ºC. Then tempered sorghum grains were poured into hot salt for popping. The salt-*bhatti* was the very good for good quality pops preparation from pop sorghum. The popping percentage on the basis of grain number and weight basis were calculated. The new genotype RPOSV-3 gave higher percentage of popping. The expansion volume was also higher than the local genotypes but less than maize (**Table 61**). The RPOSV-3 pops gave very bright white colour and extra-large, fully open size than other genotypes. The popping percentage was higher in RPOSV-3 (87.4%) and processing losses were less (3.4%) than the other genotypes studied (**Table 62**). The chemical composition of the grains and pops was found similar except having higher starch content in grains as well as in *pops* prepared from pop sorghum (**Table 63, 64**). The organoleptic score of the *pops* showed that the RPOSV-3 having similar rank to that of Maize *pops* (**Table 65**) having 8.1 overall acceptability score. The sorghum pops can be converted in to *pop* grits and flour and then use for various food items. The nutritional composition of the sorghum grain, *pop*, grits and *pops*' flour also showed very good source of minerals and vitamins (**Table 66**). The *pops*' grits and flour shoed slightly lower level of these elements due to processing losses. The antinutritional factors also reduced in the processed products (**Table 66**). The amino acid profile of sorghum grain, *pops*, *pops*' grits and *pops*' flour also showed promising results (**Table 67**). The production cost of sorghum *pops* showed that if the crop management and processing technology used properly then farmer can get nearly Rs. 1, 12, 668/- net profit from one hectare area of *pop* sorghum. The economical returns from *pop* sorghum are very good for commercialization of the *pop* sorghum (**Table 68**). Farmers also get fodder from the sorghum *pop* variety.

RPOSV-3 (Phule Panchami) having very good popping characteristics as well as nutritional quality as compared to the other local genotypes. It can be stored one year without any adverse effect in plastic bags with proper sealing. The economical returns from *pop* sorghum are very good for commercialization of the *pop* sorghum.

Table 61: Popping quality of popped *rabi* sorghum variety RPOSV 3 (Phule Panchami)*.

Popping quality parameters	Genotypes				
	RPOSV 3	**Local 1**	**Local 2**	**M 35-1**	**Maize**
Popping (%) on gain number basis	98	82	63	59	98
Expansion volume (ml/g)	6.5	4.8	4.6	2.6	19.6
Colour of pop grains	White	White	White	White	Creamy white
Size of popped grains	Extra-large white fully opened & excellent	Large white	Small whitish	Small whitish, yellow	Extra-large, fully opened, creamy white, excellent

*Results are means of 10 determinations.

Table 62: Popping percentage (on wt. basis) of various *rabi* sorghum genotypes*.

Genotype	Popped Grains %	Un-popped grains%	Processing losses%	Grain price (Rs./Kg)
RPOSV 3	87.4	8.8	3.8	20
Local 1	73.2	18.6	8.2	20
Local 2	56.6	36.4	7.0	20
M 35-1	38.0	53.8	8.2	18
Maize	82.6	4.0	13.4	52
Range	38.0-87.4	4.0-53.8	3.8-13.4	-
Mean	67.56	24.32	8.12	-
S. E. ±	18.14	18.45	3.09	-
C. D. at 5% (n=3)	54.42	55.34	9.28	-

*Results are means of three determinations.

Table 63: Chemical constituents of sorghum pop grains (RPOSV-3; Phule Panchami)*.

Parameter	RPOSV 3	Local 1	Local 2	M 35-1	Maize	S.E.±	CD at 5%
Starch (%)	75.1	70.1	74.6	64.4	69.7	3.88	11.66
Protein (%)	8.65	8.15	7.03	10.47	9.03	1.12	3.38
Soluble protein (%)	1.00	1.13	0.93	1.43	0.59	0.27	0.82
Total sugars (%)	1.25	1.11	0.87	1.92	0.93	0.37	1.13
Free amino acids (mg/100g)	48.70	51.36	38.13	83.40	23.67	19.74	59.23
Hectoliter wt. (Kg/hl)	81.11	79.36	77.76	79.21	142.08	25.11	75.33

*Results are means of three determinations.

Table 64: Nutritional constituents of *pops* prepared from *pop* sorghum variety RPOSV-3*.

Parameter	Starch (%)	Protein (%)	Soluble protein (%)	Total sugars (%)	Free amino acids (mg/100g)
RPOSV 3	75.8	8.52	0.85	1.05	23.17
Local 1	70.9	8.01	1.01	0.95	21.16
Local 2	75.1	6.93	0.86	0.71	13.05
M 35-1	65.0	10.23	1.33	1.70	33.11
Maize	70.3	9.00	0.55	0.78	13.12
Range	65.0-75.8	8.01-10.23	0.55-1.33	0.78-1.70	13.05-33.11
Mean	71.42	8.54	0.92	1.04	20.72
S. E. ±	3.88	1.09	0.25	0.35	7.43
C. D. at 5%	11.66	3.27	0.76	1.06	22.30

*Results are means of three determinations.

Table 65: Organoleptic evaluation of sorghum pops RPOSV-3 (Phule Panchami)*.

Genotype	Colour & appearance	Flavour	Crispness	Taste	Overall acceptability	DMR rank
RPOSV 3	8.6	7.4	8.0	8.4	8.1	2
Local 1	7.4	7.2	6.8	7.6	7.3	3
Local 2	6.4	6.8	6.6	7.4	6.8	4

Genotype	Colour & appearance	Flavour	Crispness	Taste	Overall acceptability	DMR rank
M 35-1	6.2	7.0	5.6	6.8	6.4	5
Maize	7.8	8.0	8.2	8.6	8.2	1
Range	6.2-8.6	6.8-8.0	6.6-8.2	6.8-8.6	6.4-8.2	-
Mean	7.28	7.28	7.04	7.76	7.36	-
S.E. ±	0.89	0.41	0.95	0.66	0.70	-
C.D. at 5%	2.67	1.24	2.87	1.99	2.12	-

*Ten semi-trained judges were used for the sorghum *pops* organoleptic/sensory quality evaluation and 1 to 9 point hedonic scale used for scoring. DMR = Duncan multiple range rank.

Table 66: Nutritional composition of sorghum grains (Phule Panchami), pops, pop grits and pops' flour*.

Parameter	Grain	Pops	Pops' grits	Pops' flour	SE ±	CD at 5%
Crude protein,%	8.65	9.04	8.37	10.82	0.95	2.86
Soluble proteins,%	1.00	1.59	1.20	1.15	0.21	0.65
Lipids,%	1.52	1.17	1.21	0.85	0.23	0.71
Crude fiber,%	3.94	2.05	2.67	3.12	0.68	2.07
Total sugars,%	1.25	1.03	1.15	1.05	0.08	0.26
Starch,%	75.01	75.26	67.35	66.33	4.18	12.56
Ash,%	2.86	2.72	2.64	2.53	0.12	0.36
Mineral elements						
Calcium, mg/100g	14.51	25.81	27.29	26.95	5.29	15.90
Potassium, mg/100g	517.57	254.35	215.53	197.98	129.33	388.01
Phosphorus, mg/100g	524.78	510.81	519.15	516.73	5.01	15.04
Sodium, mg/100g	29.93	13.11	11.90	8.61	8.27	24.82
Iron, mg/100g	9.49	6.48	5.31	7.76	1.55	4.66
Magnesium, mg/100g	219.58	217.02	207.13	211.87	4.79	14.38
Zinc, mg/100g	4.28	4.19	4.00	4.24	0.10	0.32
Copper, mg/100g	0.84	0.76	0.75	0.79	0.03	0.11
Manganese, mg/100g	2.05	2.07	2.07	1.81	0.11	0.33

Parameter	Grain	Pops	Pops' grits	Pops' flour	SE ±	CD at 5%
Vitamins						
β-Carotene, μg/100g	52.75	3.78	-	-	-	-
Thiamine, mg/100g	0.45	-	-	-	-	-
Riboflavin, mg/100g	0.32	0.34	0.36	0.35	0.01	0.04
Niacin, mg/100g	1.63	-	-	-	-	-
Folic acid, μg /100g	31.26	9.54	9.29	9.77	9.40	28.23
Antinutritional factors						
Free amino acids, mg/100g	93.09	89.04	85.70	87.50	2.72	8.18
Phenolics, g/100g	1.72	1.65	1.17	1.65	0.21	0.66
Oxalic acid, mg/100g	14.48	-	-	-	-	-
Phytic phosphorus, mg/100g	65.45	39.79	42.16	42.68	10.40	31.23

*Results are means of three determinations. - Not detected

Table 67: Amino acid composition of sorghum grains (Phule Panchami), *pops*, *pop* grits and *pops*' flour (g/16 g Nitrogen)*.

Parameter	Grain	*Pops*	*Pops*' grits	*Pops*' flour	S.E. ±	CD at 5%
Alanine	7.15	6.44	6.10	5.39	0.63	1.90
Arginine	4.89	3.30	4.72	3.73	0.35	1.08
Aspartic acid	8.72	6.07	5.33	7.58	1.31	3.95
Cysteine	1.22	0.92	0.91	0.77	0.16	0.49
Glutamic acid	20.11	23.28	24.45	24.89	1.86	5.60
Glycine	3.89	3.56	3.39	2.44	0.53	1.62
Histidine	1.89	2.49	2.77	2.72	0.34	1.05
Isoleucine	3.23	3.95	4.04	4.73	0.53	1.59
Leucine	11.86	15.46	16.72	17.86	2.25	6.76
Lysine	2.75	1.71	1.63	1.66	0.46	1.41
Methionine	1.49	1.58	1.63	1.72	0.08	0.25
Phenylalanine	4.32	4.67	4.58	5.18	0.31	0.94
Proline	5.28	5.30	5.34	3.96	0.58	1.75

Parameter	Grain	*Pops*	*Pops'* grits	*Pops'* flour	S.E. ±	CD at 5%
Serine	4.24	4.09	3.89	3.66	0.21	0.65
Threonine	3.19	3.33	3.42	3.58	0.14	0.42
Tryptophan	1.05	0.77	0.65	0.51	0.19	0.60
Tyrosine	3.48	4.57	4.91	4.57	0.53	1.62
Valine	2.46	2.40	2.33	2.39	0.04	0.14

*Results are means of three determinations.

Table 68: Economical returns from *pop* sorghum.

Variety	Grain yield Kg/ha	Net grain yield used for popping (kg/ha)	Market rate Rs/Kg	Total price of grain (Rs)	Processin g cost (Rs)	Cost of grain with processing (Rs)	Market price of processed product (Rs)	Net expected return (Rs)
RPOSV 3	1376	1202	18/-	24768/-	18824/-	43592/-	15,6260/-	1,12,668/-
M 35-1	1712	-	20/-	34240/-	-	34240/-	34240/-	Nil

Note:

1. Processing cost 76% on raw material cost: (Ref: Bakery products book, 2008; J.K. Chavan and V.A. Dhotre)
2. Sorghum pop recovery from 1 kg grain produces 26 packets of 500 ml/ 25 g each.
3. Present market price of sorghum pop packet is Rs. 5/each (i.e. Rs. 130/kg grain)
4. Un-popped grains 8.8%, popping losses 3.8% (i.e. total loss is 12. 6%).

Sorghum pop laddu

First prepare sorghum pops as mentioned in the preparation of sorghum pops. The good quality sorghum pops then used for the preparation of sorghum pop laddu.

Ingredient	Quantity
Sorghum pops	1000 g
Jaggary/Sugar	600 g
Water	250 g

Procedure for the preparation of sorghum pop laddu

Dissolve sugar/jaggery in warm water and then heat on medium flame until it get sticky consistency and then add sorghum pops and mix it very well. All pops must be sticky with jaggery/sugar syrup. While it is warm taking small quantity and press with two hands in such a way that it forms a round shape laddu and keep for cooling and hardening. Then pack into plastic bags or plastic container and store in dry clean place. It can remain in good condition for about 15 days.

Sorghum pop grits laddu

First prepare sorghum pops as mentioned under preparation of sorghum pops. Then prepare sorghum pops grits using mixer on small scale for large scale grinding mill can be used.

Ingredient	Quantity
Sorghum pop grits	1000 g
Jaggary/Sugar	600 g
Water	250 g

Procedure for the preparation of sorghum pop grits laddu

Dissolve sugar/jaggery in warm water and then heat on medium flame until it get sticky consistency and then add sorghum pops grits and mix it very well. All pops grits must be sticky with jaggery/sugar syrup. While it is warm taking small quantity and press with two hands in such a way that it forms a round shape laddu and keep for cooling and hardening. Then pack into plastic bags or plastic container and store in dry clean place. It can remain in good condition for about 15 days.

Sorghum pop grits *chikki*

First prepare sorghum pops as mentioned under preparation of sorghum pops. Then prepare sorghum pops grits using mixer on small scale, for large scale grinding mill can be used.

Ingredient	Quantity
Sorghum pop grits	1000 g
Jaggary/Sugar	600 g
Fat	30 g
Water	250 g

Procedure for the preparation of sorghum pop grits *chikki*

Take water in a frying pan, warm it and then add jaggery/sugar, heat till it becomes like a soft paste. Add sorghum pop grits, mix well and then transfer the mixture to a greased flat plate and spread evenly. Allow to cool and then cut into attractive pieces. Store in airtight container. It can remain more 30 days in good condition at room temperature.

Sorghum pop flour *chikki*

Prepare sorghum pops as mentioned under sorghum pops preparation. Then grind the sorghum pops into fine powder using grinder/mixer or grinding mill. Then use this sorghum pop flour (80 mesh) for preparation of chikki.

Ingredient	Quantity
Sorghum pop flour	1000 g
Jaggary/Sugar	600 g
Fat	30 g
Cardamom powder	1 tea spoon
Water	250 g

Procedure for the preparation of sorghum pop flour *chikki*

Take water in a frying pan, warm it and then add jaggery/sugar heat till it becomes like a soft paste. Then add sorghum pop flour, cardamom powder, mix well and then transfer the mixture to a greased flat plate and spread evenly. Allow to cool and then cut into attractive pieces. Then store in plastic bags or containers in dry place until use.

Sorghum pop flour health drink

First prepare sorghum pops. Then grind sorghum pops in to fine powder using mixer or grinder at home or laboratory scale but for large-scale use grinding mill. Sieve the sorghum pop flour using 80-mesh sieve.

Ingredient	Quantity
Sorghum pop flour	10 g
Badam powder	5 g
Sugar	5%
Water/Milk	100 ml
Cocoa powder for flavour	As per requirement

Procedure for the preparation of sorghum pop flour health drink

Ground sugar and badam in to fine powder. Mix it into sorghum pop flour and then add cocoa powder. This all mixture adds in to warm/cold water or milk and homogenate properly and used as a health drink. It is good health drink for children as well as for adults.

10. Sorghum Flakes (Pohe)

Procedure for the preparation of sorghum *Flakes (Pohe)*

First clean and wash the sorghum grains thoroughly. Soak the grains in hot water for about 24 hrs. Drain the water and place sorghum grains in a cooker and steam at 15 Psi for 10 – 15 min. Take it out and pass through flaking machine and dry it under shed. This sorghum flakes/*Poha* can be used in sorghum *Chiwada* preparation or onion-*Poha* preparation. This *Poha* can be preserve by placing in LDPE bags by vacuum packaging. It can be stored upto 2 – 3 months but it is not suitable for storage during winter season. The nutritive value of sorghum Flakes/*Poha* is given in **Table 60**.

Sorghum low fiber flour

Procedure for the preparation of sorghum low fiber flour

Clean sorghum grains properly. Grind it into fine flour using grain-grinding mill. Sieve the flour through 60-mesh sieve and separate coarse material collected on top of the sieve, and then we get low fiber flour. Such type of flour can be used for the preparation of various products. The nutritive value of sorghum low fiber flour is given in **Table 60**.

Sorghum high fiber flour

Procedure for the preparation of sorghum high fiber flour

Clean sorghum grains properly. Grind it into fine flour using grain-grinding mill. Sieve the flour through 80 or 100-mesh sieve and separate small quantity fine flour then we get remaining high fiber flour. Such type of flour can be used for the preparation of specific products. The nutritive value of sorghum high fiber flour is given in **Table 60**.

Sorghum malt flour

Procedure for the preparation of sorghum malt flour

First clean and wash sorghum grains properly. Steep the grains in warm water overnight. Drain water from the grains and wash thoroughly. Place the steeped grains in muslin cloth and tie loosely keeps for the germination at warm place upto 30–36 hrs. After germination dry all grains and remove the sprouts and grind into fine flour. Sieve the flour through 60-mesh sieve and separate fine flour and fiber. The fine malted sorghum flour can be used for preparation of various ethnic and health foods. It can be packed in LDPE bags and preserve by vacuum packaging. The nutritive value of sorghum malt flour is given in **Table 60**.

Sorghum diabetic flour

Procedure for the preparation of sorghum malt flour

Take cleaned sorghum grains and grind it into fine flour. Add 20% malt sorghum flour, 5% dehydrated mushroom powder, 2% dehydrated carrot, 5% malted soybean flour, 5% fenugreek leaves powder, 3% bitter gourd powder and 8% malted finger millet whole flour. Sieve mixture of all flours through 20-mesh sieve. Then use such type of flour for preparation of various diabetic foods. This type of flour can be used for the preparation of *Bhakari* or *Thalipeeth*. It can be preserve in LDPE bags by vacuum packaging for short time. The nutritive value of sorghum diabetic flour is given in **Table 60**.

Sorghum *semolina/Rava*

Procedure for the preparation of sorghum *semolina/Rava*

First clean sorghum grains wash it thoroughly. Steep it into hot water for 24 hrs. After that drain the water properly and place the grains in pressure cooker and steam under 15 Psi for 10-15 min. Then remove it from pressure cooker and keep for drying to reduce moisture upto 12%. After that grind sorghum grains in grain Mill coarsely. Sieve it through using 40-mesh sieve to obtain sorghum Semolina/ *Rava* grits and flour. Such type of semolina can be used for the preparation of *Idli* or *Dosa*. It can be preserved in LDPE pouch by vacuum packaging and it remains in good condition for about 6 months. The nutritive value of sorghum semolina/*Rava* is given in **Table 60**.

Sorghum *Noodles*

Ingredient	Quantity
Pearled sorghum grain flour	1000 g
Sorghum semolina	250 g
Salt	As per taste
Water	As required to wet the flour

Procedure for the preparation of sorghum *Noodles*

First sieve the sorghum flour and semolina. Then mix with water and knead in to smooth soft dough. After that cook this dough for 5 min in a pressure cooker. Then keep this cooked dough in a pressing machine (*Sorya*) or noodles making machine and press the noodles on a polyethylene paper or sheet. Dry the noodles under sun and then store in the plastic bags for at least 6 months without any disorder. The nutritive value of the sorghum noodles is given in the Table 60.

Kheer from sorghum *noodles*

Ingredient	Quantity
Sorghum noodles	500 g
Cashew nuts	200 g
Milk	2.5 Liters
Raisins	50 g
Sugar	500 g
Fat	50 g
Salt	To taste
Elachi	10 No
Water	As per requirement

Procedure for the preparation of sorghum *Noodle Kheer*

First fry the noodles with fat on oven separately. Then fry dry fruits with fat in a pan. After that boil the water and milk together then add fried noodles. Cook for some time and then add sugar, mix slowly and cook for 10 – 15 min. At the last add elachi powder and decorate with cashew nuts, raisins and other dry fruits. Serve when it is hot. The nutritive value of the sorghum *noodle kheer* is given in the Table 60.

Sorghum *Modak*

Ingredient	Quantity
Sorghum *Semolina/Rava*	200 g
Sorghum flour	200 g
Sugar	100 g
Fat	50 g
Salt	To taste
Water	As per requirement

Procedure for the preparation of sorghum *Modak*

First clean the sorghum *semolina/rava*. Boil the water in a container and add *rava* slowly and cook for few minutes. Then add sorghum flour, sugar, salt and fat in it while mixing simultaneously. When half cooked remove from the flame and cool it, make round balls and then cook like normal *modak* in a pressure cooker or on sieve on hot water container for 10 – 15 min. Cool the modak at room temperature and then serve it as a sweet product. This product can be prepared without sugar and it can serve with tomato/tamarind soup or samber/curry. The nutritive value of the sorghum *modak* is given in the Table 60.

Traditional and Ethnic Sorghum Products

Sorghum *Dashmi*

Ingredient	Quantity
Sorghum flour	120 g
Wheat flour	80g
Sugar	40 g
Fat/Oil	30 g
Salt	As per taste
Cardamom	6 g
Milk	200 g

Procedure for the preparation of sorghum *Dashmi*

Sorghum and wheat flour sieve together. Add sugar, salt, cardamom powder in milk stir well. Add milk in sorghum and wheat flour, mix well and knead into smooth dough by adding little oil. After that make small ball and spread it like chapatti in round shape. Roast on pre heated iron sheet/Tawa. Apply small quantity of oil on it and roast till it become light brown. Serve it immediately as hot or preserve in airtight LDPE bag for further use. The nutritive value of sorghum *Dashmi* is given in **Table 60**.

Sorghum *Khakara*

Ingredient	Quantity
Sorghum flour	100 g
Wheat flour	If required small quantity
Oil	As per requirement
Salt	As per taste
Turmeric powder	Pinch
Chilli powder	To taste
Water	As per requirement

Procedure for the preparation of sorghum *khakara*

Take sieved sorghum flour, for stickiness adds small quantity of wheat flour. Mix other all ingredients and make dough. Roll it into very thin sheet. Some times this thin sheet can be cut into small strips using a knife. Then shallow fry, till it become brown and crunchy. Such type of product can be stored for at least one week in good condition at room temperature. The nutritive value of sorghum *Khakara* is given in **Table 60**.

Sorghum *Thalipeeth*

Ingredient	Quantity
Plain or malted or *hurda* sorghum flour	150 g
Soybean flour	50 g
Salt	as per taste
Oil	40 ml
Onion (chopped)	100 g
Green chilli (chopped)	to taste
Coriander leaves	20 g
Garlic	to taste
Ginger	to taste
Water	as per requirement

Procedure for the preparation of sorghum *Thalipeeth*

Sieve all flours with sorghum flour using 60-mesh sieve. Cut onion, green chilli, coriander leaves, garlic, ginger in to fine pieces. Make garlic, ginger and chilli paste. Mix sorghum flour and other remaining all ingredients together. Make soft dough with required quantity of water and knead into smooth dough. Take small quantity of dough, make small ball and spread it in a round shape with the help of hand on cloth or polythene paper into round shape. After that place on shallow fry pan or pre-heated Tawa and bake on both sides by applying little oil on it. Serve it as hot with chutney or curd. It can remain in good condition upto one day at room temperature. The nutritive value of sorghum *Thalipeeth* is given in **Table 60**.

Sorghum *papad*

Papad is a snack food item and it is very tasty so it is used in every Indian's diet sine old day. People like this item very much so it has great demand in every corner of the country. A variety of *papads* are prepared from pulses *dhals* and cereals. Basically papads are prepared from black gram dhal, *mung dhal* etc. In

South India, *papads* are familiar as *applams*, whereas in Gujarat and Rajasthan as *Khachche*. *Papads* are traditionally prepared by hand rolling methods employing ladies at every house. Papad is a small flat *Chapati* like structure made from *dhal* flour which is kneaded into a stiff dough after mixing with spices, salt, oil and leavening agents. Further, small uniform dough balls are prepared and rolled into flat shape like *chapati* into round shape by hand rolling. Then they are dried upto desirable moisture level and packed into LDPE bags. *Papads* can be classified on the basis of their size and shape as well as ingredients contents in it. Now a days *papads* are produced by using *papad* machines. The malted sorghum flour with malted *ragi* flour produce excellent nutritious papads, having high fiber. Sorghum with other pulses and cereals produce nutritionally superior *papads* than the traditional *papads*. The general procedure of malting of grains is given **Figure 15**.

Sorghum / Ragi /Bajara
↓
Grains
↓
Cleaning and washing
↓
Steeping in hot water (overnight)
↓
Drain water and wash thoroughly
↓
Keep in cloth for germination (40 –48 hrs)
↓
Dry at low temperature
↓
Remove germs
↓
Grind into fine flour
↓
Sieving (60-mesh sieve)
↓
Fine malted flour
↓
Use for various types of products

Figure 15: Flow sheet for preparation of malted flour from various crop grains.

A. Malted sorghum *papad*

Ingredient	Quantity
Malted sorghum flour	500 g
Papadkhar	5 g
Salt	20 g
Gingelly seed	5 g
Asafoetida	2.5 g
Cumin seed powder	5 g
Black pepper powder	2.5 g
Water	500 ml

B. High fiber mixed *papad*

Ingredient	Quantity
Malted sorghum flour	375 g
Malted *Ragi* bran	125 g
Papadkhar	5 g
Salt	20 g
Sesamum	5 g
Asafoetida	2.5 g
Cumin seed powder	5 g
Black pepper powder	2.5 g
Ajwan	2.5 g
Water	500 ml

C. Malted mixed sorghum *papad*

Ingredient	Quantity
Malted sorghum flour	500 g
Malted *Ragi* flour	500 g
Papadkhar	10 g
Salt	36 g
Gingelly seed	10 g
Asafoetida	2.5 g
Cumin seed powder	5 g
Black pepper powder	5 g
Water	1000 ml

D. Chilli *papad*

Ingredient	Quantity
Malted sorghum flour	150 g
Malted *Ragi* bran	100 g
Papadkhar	3 g
Salt	12 g
Chilli powder	3 g
Garlic paste	5 g
Gingelly seed	5 g
Asafoetida	2.5 g
Cumin seed powder	5 g
Black pepper powder	2.5 g
Water	450 ml

Procedure for the preparation of all types sorghum *papads*

First sieve all flours together. Boil water and add *papadkhar*, salt, sesame seeds, asafetida, cumin powder and black pepper powder and all other spices. Then add flour mixture in boiling water, stir well and mix into smooth dough. Divide the dough into small portions 10 – 12 g each ball. Steam the balls in steamer or pressure cooker at 15 Psi for 10 – 15 min. Place the balls in *papad* press machine after enclosing in plastic sheet and press into 0.10 or 0.15 cm thick sheet. Dry at room temperature mostly under shed until get desired moisture. Pack in laminated bags and store at room temperature. It can remain in good condition upto one year without any spoilage. The nutritive value of sorghum *papads* is given in **Table 60**.

Sorghum *Upama*

Upama is generally used as breakfast food and some time as a snack food too. *Upama* mostly prepared using wheat semolina and black gram dhal. But we can prepare *Upama* using sorghum semolina also.

Ingredient	Quantity
Sorghum semolina (*Rava*)	100 g
Groundnut	10 g
Black gram	25 g
Salt	To taste

Ingredient	Quantity
Sugar	To taste
Green chilli (chopped)	1.5 g
Turmeric powder	1.5 g
Cumin seed	1.5 g
Mustard seed	1.5 g
Asafoetida	2.5 g
Coriander leaves	Few
Curry leaves	Few
Oil	20 ml
Onion (chopped)	50 g
Water	As per requirement

Procedure for the preparation of all types sorghum *Upama*

Sieve sorghum semolina and roast it, till it becomes light brown. Soak black gram dhal in hot water for 8 – 10 min. Cut onion, chilli, coriander leaves, curry leaves. Place deep fry pan on gas and add oil heat for to warm, add mustard, cumin seed, asafetida, green chilli, curry leaves, groundnut, onion, black gram dhal, turmeric and fry thoroughly and after that add sufficient water in it, boil it and add sorghum semolina, sugar, salt etc. After cooking spread coriander leaves on it and serve as hot by garnishing with coconut shreds and lime piece. The nutritive value of sorghum *Upama* is given in **Table 60**.

Sorghum *Pakoda* (*Bhagi*)

Ingredient	Quantity
Sorghum flour	200 g
Chickpea flour	100 g
Ginger paste	1 tea spoon
Salt	To taste
Chilli powder	To taste
Onion (chopped)	2
Turmeric powder	1 pinch
Green chillies (chopped)	4
Coriander leaves	1 bunch
Curry leaves	1 bunch
Oil	400 ml
Water	As per requirement

Procedure for the preparation of all types sorghum *Pakoda (Bhagi)*

Sieve sorghum flour and chickpea flour together. Add salt, chilli powder, chopped onion, chopped green chillies, chopped coriander leaves, ginger paste to the sieved flour and mix with hot water and make smooth dough consistency. Heat the oil and put this dough in different shapes, deep fry using medium flame till it attains golden colour. Serve hot with tomato sauce. The nutritive value of sorghum *Pakoda (Bhagi)* is given in **Table 60**.

Sorghum *Shira* (sweets)

Shira is a sweet product generally used during breakfast. This product is rich in protein, fat and minerals. This product also used for snack to the guests and relatives while visiting as well as to the patients to get more energy.

Ingredient	Quantity
Sorghum semolina (*Rava*)	100 g
Groundnut/Cashew nuts/Raisins	50 g
Sugar	100 g
Elichi fruits	5 g
Salt	To taste
Milk	100 ml
Fat	50 g
Keshari colour	1 pinch
Water	As per requirement

Procedure for the preparation of all types sorghum *Shira* (sweets)

Sieve sorghum semolina and roast it, till it becomes light brown. Place deep fry pan on gas and add oil heat for to warm, add sufficient milk and water in it, boil it and add sorghum semolina, sugar, salt etc. After cooking spread chew nuts/groundnuts/raisins etc. and serve it while hot. The nutritive value of sorghum *Shira* (sweets) is given in **Table 60**.

Sorghum *Rava Idali*

Ingredient	Quantity
Parboiled sorghum semolina (*Rava*)	100 g
Rice	50 g
Salt	To taste
Black gram	50 g
Curd	200 g
Coriander	10 g
Water	As per requirement

Procedure for the preparation of all types sorghum *Rava Idali*

First soak rice and paste it, simultaneously soak the black gram dhal for 8 hrs and drain out the water grind into fine paste. Mix these pastes and ferment overnight. Soak parboiled *Rava* for 1.5 hrs. Add salt, soaked Rava in fermented paste and mix it thoroughly with curd and coriander leaves. Apply teaspoon oil to the *idli* mould and pour *idli* batter in it. Steam it in *idli* cooker for 10 –15 mins, take out and remove *idli* from the moulds and serve hot with chutney. The nutritive value of sorghum *Idali* is given in Table 60.

Sorghum *dosa*

Ingredient	Quantity
Sorghum semolina (*Rava*)	80 g
Rice	100 g
Salt	To taste
Black gram	70 g
Oil	200 g
Curd	50 g
Coriander leaves	Few
Colour	As per requirement
Water	As per requirement

Procedure for the preparation of all types sorghum *dosa*

First soak rice and black gram dhal for 8 hrs and drain out the water grind into fine paste. Ferment the batter overnight. Soak parboiled *Rava* for 0.5 hrs. and add salt, soaked *Rava* in fermented paste and mix it thoroughly with curd and coriander leaves. Apply teaspoon oil on the pre heated *Dosa* making *Tawa* and pour dosa batter on it, smear it by round scoop into thin-round shape. Fry it in shallow fry pan/*Tawa* and serve as hot with chutney. The nutritive value of sorghum *dosa* is given in Table 60.

Sorghum *Uttappa*

Ingredient	Quantity
Parboiled/plain sorghum semolina (*Rava*)/Flour	80 g
Rice	100 g
Salt	To taste

Ingredient	Quantity
Black gram	70 g
Oil	150 g
Onion	200 g
Tomato	200 g
Green chilli	20 g
Coriander leaves	10 g
Water	As per requirement

Procedure for the preparation of all types sorghum *Uttappa*

First soak rice and black gram dhal for 8 hrs and drain out the water, grind into fine paste and add sorghum *Rava* in it. Ferment the batter overnight. Mix it and make it into thin batter consistency by adding required quantity of water. Cut onion, green chilli, coriander leaves, tomato and add in fermented batter. Add salt, and mix well. Put little oil on dosa pan, pour the batter on it and spread the batter round and shallow fry. Serve as hot with chutney or any type of curry. The nutritive value of sorghum *Uttappa* is given in Table 60.

Sorghum *Appe*

Ingredient	Quantity
Malted sorghum semolina (*Rava*)	100 g
Rice	50 g
Salt	To taste
Black gram dhal	100 g
Oil	20 g
Cumin seed powder	2 g
Curd	200 ml
Water	As per requirement

Procedure for the preparation of all types sorghum *Appe*

First soak rice and black gram dhal for 6 hrs and drain out the water, grind into fine paste and Ferment overnight. Soak sorghum *Rava* for 0.5 hr mix with curd, sorghum *Rava*, salt, and cumin seed powder into fermented the batter. Apply little oil on *Appe* stand and pour the batter in it. Steam for 10 min. Remove from the Appe stand and serve as hot with chutney. The nutritive value of sorghum *Appe* is given in Table 60.

Sorghum *Chiwada*

Ingredient	Quantity
Sorghum flakes	100 g
Groundnut	15 g
Coconut (desiccated/cut)	10 g
Salt	To taste
Oil	10 g
Sugar powder	To taste
Chilli powder	To taste
Turmeric powder	1 Pinch
Cumin seed	2 g
Mustard seed	2 g
Coriander leaves	10 g
Curry leaves	Few
Chew nuts	10 g
Raisins	10 g

Procedure for the preparation of all types sorghum *Chiwada*

Fry sorghum flakes in sufficient quantity of oil. Fry coconut and groundnut. Cut coriander leaves. Add all ingredients in oil and fry well. Add fried flakes in it and mix well. Serve hot or preserve by packing into LDPE bags. For longer storage apply nitrogen gas flush and then seal the pouch. The nutritive value of sorghum *Chiwada* is given in Table 60.

Sorghum *Rava laddu*

Ingredient	Quantity
Sorghum *Rava* (semolina)	100 g
Soybean flour	50 g
Sugar powder	40 g
Salt	To taste
Fat	30 g
Almond	8 g
Raisins	8 g
Chew nuts	8 g
Cardamom	1 pinch

Procedure for the preparation of all types sorghum *Rava laddu*

Fry sieve sorghum *Rava* and soybean flour together. Roast the mixture adding fat till it get good smell and light brown colour. Roast chew nuts, almond and raisins in little fat. Add sugar powder chew nuts, almonds, raisins and cardamom powder in the roasted material and mix it well. Divide the mixture into small equal portions and prepare *laddu*. Serve or store for few days in plastic bags or stainless steel container. The nutritive value of sorghum *Rava laddu* is given in Table 60.

Sorghum *Chakali*

Ingredient	Quantity
Sorghum grains	100 g
Wheat	50 g
Rice	50 g
Salt	To taste
Oil	200 g
Green gram dhal	50 g
Bengal gram dhal	50 g
Black gram dhal	25 g
Soybean	25 g
Gingelly seed	15 g
Cumin seed powder	5 g
Ajwan seed	5 g
Chilli powder	To taste
Boiled water	As per requirement

Procedure for the preparation of all types sorghum *Chakali*

Clean, wash, dry all the grains and pulses separately. Roast all separately, mixes and grind into fine flour. Sieve flour and add chilli powder, cumin and ajwan seed powder, gingelly seed, salt and boiled oil and mix well into smooth dough. Add boiled water and make smooth dough. Place it into *Chakali* making *Sorya*, press it and put into hot oil for frying. Deep fry in oil using low flame, till it attains golden yellow colour. Serve after cooling and store in plastic bags or in stainless steel container for few days. The nutritive value of sorghum *Chakali* is given in **Table 60**.

Sorghum *Shev*

Ingredient	Quantity
Sorghum flour/malted sorghum flour	100 g
Wheat flour	20 g
Soybean flour	20 g
Cumin seed powder	5 g
Ajwan seed powder	5 g
Chilli powder	10 g
Oil	250 ml
Salt	To taste
Boiled water	As per requirement

Procedure for the preparation of sorghum *Shev*

Mix all flours and then sieve. Then add chilli powder, cumin and ajwan seed powder, salt and small quantity of oil and mix well. Add boiled water and make smooth dough. Place it in shev maker (Sorya) press it in too deep frying oil using low flame, till it attains golden colour.

Sorghum *Kurdai*

Ingredient	Quantity
Sorghum grains	1000 g
Salt	To taste
Water	As per requirement

Procedure for the preparation of all types sorghum *Kurdai*

First clean and wash sorghum grains properly. Deep sorghum seeds in excess water for 3 days. Change water every day. After that wash thoroughly and grind in mixer. Make fine paste. Strain the paste by sieve or muslin cloth. Keep the strained solution aside for one night to settle slurry. After 12 – 15 hrs, remove all water from slurry. Take that slurry and add in sufficient boiling water. After that add salt to taste in it and cook it for 5 –10 min. till to becomes semi- solid or thick slurry. Put it in *Kurdai* press machine/*Sory* and make *Kurdai* into round shape threads. Dry it in sunlight sufficiently. When we want to use it fry in boiling oil. For future use store it in clean container at room temperature. It can remain in good condition upto one year. The nutritive value of sorghum *Kurdai* is given in **Table 60**.

Sorghum *Papadi*

Ingredient	Quantity
Sorghum grains	1000 g
Crushed cumin seed	5 g
Salt	To taste
Colour	As per requirement
Water	As per requirement

Procedure for the preparation of all types sorghum *Papadi*

First clean and wash sorghum grains properly. Deep sorghum seeds in excess water for 3 days. Change water every day. After that wash thoroughly and grind in mixer. Make fine paste/slurry. Strain the paste by sieve or muslin cloth. Keep the strained solution aside for one night to settle slurry. After 12 – 15 hrs, remove all water from slurry. Take that slurry and add salt, crushed cumin seed in it and mix it well. Put one tablespoon of batter on a plain plate and rotate it till we get round shape of batter/*Papadi*. Put this plate for steaming on boiling water for 3 – 5 min. Remove *Papadi* with the help of needle and dry sufficiently under shed. Serve after frying gives crunchy-n-tasty. The nutritive value of sorghum *Papadi* is given in **Table 60**.

Sorghum *Vada*

Ingredient	Quantity
Sorghum grains	1000 g
Cumin seed powder	5 g
Salt	To taste
Chilli powder	To taste
Water	As per requirement

Procedure for the preparation of sorghum *Vada*

First clean and wash sorghum grains properly. Deep sorghum seeds in excess water for 3 days. Change water every day. After that wash thoroughly and grind in mixer. Make fine paste/slurry. Strain the paste by sieve or muslin cloth. Keep the strained solution aside for one night to settle slurry. After 12 – 15 hrs, remove all water from slurry. Take that slurry and add cumin seed powder and salt in it, cook for 5 – 10 min. till it becomes solid. Make small balls and dry it in sunlight sufficiently. Serve after frying in oil. The nutritive value of sorghum *Vada* is given in **Table 60**.

Health drink from sorghum

Ingredients	Quantity
Sorghum grains	3 kg
Ragi grains	1 kg
Soybean	1 kg
Green gram	1 kg
Bajra	2 kg
Arecanuts	300 g
Sugar	150 g
Cocoa powder	As required (for flavour)
Preservative	1 g
Water	As per requirement

Procedure for preparation of health drink

First soak different types of grains in different containers for overnight. Then wash the grain with tap water and pack loosely in cloth bags, which are already wetted with, distill water and kept for germination. It may take 2 - 3 days to germinate. These germinated grains are dried in sunlight for 2-3 days. Remove the sprouts then floured finely and separately. Ground the sugar and arecanuts. Mix all different flours and add this sugar mixture to the above flour mixture and add cocoa powder for taste and add preservative to preserve the flour. Add this mixture may be 10 to 15% in warm water or in warm milk and use as a health drink. It is very good health drink for children and also for adults too.

Bakery Products from Sorghum

Sorghum sweets, biscuits and cookies

Biscuits, sweets and cookies are the most popular common snack food items used by everybody at any time. It provides better energy, proteins, minerals and vitamins with longer shelf life. It can be used for incorporating higher amount of proteins, minerals and vitamins. Malted or germinated sorghum can be used for the improvement the nutritional quality of biscuits and cookies. Malting or germination already reported for the improvement in the overall nutritional quality, digestibility of proteins and starch. The higher proportion of dietary fiber of sorghum confers many advantages such as nutritional, digestive and physiological benefits. The high fiber sorghum cookies can provide low calories with high fiber and more resistant starch, which are reported to highly useful for diabetic and obese patients. Malted sorghum flour has been reported to improve the nutritional value. The malted cookies could be specially developed for specific targeted population such as school children, diabetics, and obese people.

Sorghum sweet rolls

Ingredient	Quantity
Sorghum flour	400 g
Maida	100 g
Sugar/Jaggery	400 g
Milk	1000 ml
Elachi	5
Water	As required

Procedure for the preparation of Sorghum sweet rolls

First sieve the sorghum and wheat flours together. Mix these flours with warm water to make dough. Cook this dough in a pressure cooker for 10 min. Grate the jaggery, add some water and boil to dissolve in water. Remove from the flame. Boil the milk, put the dough in to the *sorya* or pressing machine and press it in hot milk step by step by keeping some time gap. Cook for some time and remove from the flame, cool it and mix it in cooled jaggery. Then add elachi powder, cashew nuts and dry fruits for the taste. The nutritive value of the sorghum sweet rolls is presented in **Table 60**.

Sorghum *muffins*

Ingredient	Quantity
Sorghum flour	200 g
Fat	100 g
Sugar powder	150 g
Eggs	4
Baking powder	3 g
Vanilla	2-3 drops
Sultanas	85 g

Procedure for the preparation of Sorghum *muffins*

First sieve the sorghum flour and baking powder together. Cream the fat and sugar together to get good forth. Beat the eggs thoroughly and mix few drops of essence. Then add beaten eggs to the rubbed fat and sugar mixture. Cut the sultanas and mix to the flour. Add this flour to the above mixture and bake at 180°C for 15 –20 minutes in an oven. This can be served as a snack food. The nutritive value of sorghum muffins is given in **Table 60**.

Sorghum *Burffi*

Ingredient	Quantity
Pearled sorghum flour	200 g
Fat	20 g
Sugar	150 g
Groundnuts	50 g
Cashew nuts/Almonds	50 g
Elachi	3
Colour (yellow/kesari)	As required
Water	As per requirement

Procedure for the preparation of Sorghum *burffi*

Sieve the sorghum flour and then fry in fat or oil. Roast the groundnuts, remove the outer skin and powder it. Make small pieces of all dry fruits. Add water to the sugar and mix and boil to get thread consistency. Remove from the flame and mix the sorghum flour, groundnut flour and elachi powder and mix well. Grease the tray and spread it all over the tray. Cut this with a knife in diamond or various shape. Serve it as a traditional or occasional sweet. The nutritive value of *burffi* is given in **Table 60**.

Sorghum Cookies

Ingredient	Quantity
Sorghum flour	250 g
Maida	250 g
Sugar powder	280 g
Fat	200 g
Salt	5 g
Dextrose solution	5 ml
Cardamom powder	5 g
Nutmeg powder	5 g
Ammonium bicarbonate	2 g
Sodium bicarbonate	5 g
Water	50 ml

Procedure for the preparation of Sorghum Cookies

First sieve sorghum flour, maida, ammonium bicarbonate, sodium bicarbonate, cardamom powder and nutmeg powder together. Then cream the fat and sugar till gets light and fluffy. Take water and salt make solution and add it in the creamed fat. Add dextrose solution and then sieved flour. Knead the mixture very well and keep for 15 min. After that roll the dough into 0.5 cm thick sheet and cut with fancy biscuit/cookies cutter. Place the cookies/biscuits on baking tray and bake at 180ºC for about 10 – 15 min. to get good colour then remove from the oven, cool it and pack in plastic bags for storage. Such types of biscuits/cookies can remain in good condition upto 10 – 15 days. The nutritive value of such type of cookies and biscuits are given in **Table 60**.

Sorghum orange biscuits

Ingredient	Quantity
Malted sorghum flour	620 g
Malted soybean flour	80 g
Maida	100 g
Sugar powder	450 g
Fat	150 g
Salt	5 g
Orange juice	25 ml
Liquid glucose	50 g
Curd	100 ml
Milk powder	30 g
Orange colour	8 ml
Orange essence	7 ml
Ammonium bicarbonate	3 g
Sodium bicarbonate	5 g

Procedure for the preparation of sorghum orange biscuits

First sieve malted sorghum, soybean flours, maida, milk powder, ammonium bicarbonate, and sodium bicarbonate together. Then cream the fat and sugar till gets light, creamy white and fluffy. Then add flour mixture in it. Take curd, water and salt make solution and add it in the creamed fat. Then add in this orange juice, mix well, and make stiff dough. Add orange colour and orange essence, mix well and stiff dough. Divide the dough into small portions or use rotary moulded biscuit machine/bake at 180 - 190ºC for about 7 – 10 min. to get good colour then remove from the oven, cool it and pack in plastic bags for storage. The nutritive value of such type of cookies and biscuits are given in **Table 60**.

Malted sorghum cookies

Ingredient	Quantity
Malted sorghum flour	400 g
Maida	100 g
Sugar powder	280 g
Fat	200 g
Salt	5 g
Dextrose solution	75 ml
Cardamom powder	5 g
Nutmeg powder	5 g
Ammonium bicarbonate	2 g
Sodium bicarbonate	6 g
Water	35 ml

Procedure for the preparation of malted sorghum Cookies

First sieve malted sorghum flour, maida, ammonium bicarbonate sodium bicarbonate, cardamom powder and nutmeg powder together. Then cream the fat and sugar till gets light and fluffy. Take water and salt make solution and add it in the creamed fat. Add dextrose solution and then malted sieved flour. Knead the mixture very well and keep for 15 min. After that roll the dough into 0.5 cm thick sheet and cut with fancy biscuit/cookies cutter. Place the cookies/biscuits on baking tray and bake at 180°C for about 10 – 15 min. to get good colour then remove from the oven, cool it and pack in plastic bags for storage. Such types of biscuits/cookies can remain in good condition upto 10 – 15 days. The nutritive value of such type of cookies and biscuits are given in **Table 60**.

Sorghum chocolate cookies

Ingredient	Quantity
Malted or plain sorghum flour	200 g
Maida	300 g
Sugar powder	280 g
Fat	200 g
Cocoa powder	15 g
Salt	5 g
Dextrose solution	75 ml

Ingredient	Quantity
Ammonium bicarbonate	2 g
Sodium bicarbonate	6 g
Water	40 ml

Procedure for the preparation of sorghum chocolate cookies

First sieve malted or plain sorghum flour, maida, cocoa powder, ammonium bicarbonate, and sodium bicarbonate together. Then cream the fat and sugar till gets light and fluffy. Take water and salt make solution and add it in the creamed fat. Add dextrose solution and then malted or plain sieved flour. Knead the mixture very well and keep for 15 min. After that roll the dough into 0.5 cm thick sheet and cut with fancy biscuit/cookies cutter. Place the cookies/biscuits on baking tray and bake at 180ºC for about 10 – 15 min. to get good colour then remove from the oven, cool it and pack in plastic bags for storage. Such types of biscuits/cookies can remain in good condition upto 10 – 15 days. The nutritive value of such type of cookies and biscuits are given in **Table 60**.

Sorghum cookies for diabetic patients

Ingredient	Quantity
Sorghum high fiber flour	400 g
Malted sorghum flour	200 g
Maida	150 g
Malted Ragi bran	50 g
Malted soybean flour	100 g
Fenugreek leaves	150 g
Cumin powder	6 g
Aspartame saccharin	20 g
Fat	250 g
Salt	8 g
Curd	250 g
Bitter gourd	50 g
Dextrose solution	50 ml
Ammonium bicarbonate	3 g
Sodium bicarbonate	6 g
Water	35 ml

Procedure for the preparation of sorghum cookies for diabetic patients

First sieve all flours, maida, ammonium bicarbonate, and sodium bicarbonate together. Then cream the fat and sugar till gets light, creamy white and fluffy. Add cleaned fenugreek leaves paste, bitter gourd paste, cumin seed powder, and then mix well. Take curd, water and salt make solution and add it in the creamed fat. Add dextrose solution and then sieved flour. Knead the mixture very well. After that roll the dough into 0.5 cm thick sheet and cut with fancy biscuit/ cookies cutter. Place the cookies/biscuits on baking tray and bake at 180°C for about 8 – 10 min. to get good colour then remove from the oven, cool it and pack in plastic bags for storage. Such types of biscuits/cookies can remain in good condition upto 10 – 15 days. The nutritive value of such type of cookies and biscuits are given in **Table 60.**

Sorghum *nankhatai*

Ingredient	Quantity
Malted sorghum flour	300 g
Maida	200 g
Sugar powder	280 g
Cardamom powder	10 g
Fat	150 g
Salt	5 g
Curd	100 g
Dextrose solution	75 ml
Ammonium bicarbonate	2 g
Sodium bicarbonate	5 g

Procedure for the preparation of sorghum *nankhatai*

First sieve malted sorghum flour, maida, cardamom powder, ammonium bicarbonate, and sodium bicarbonate together. Then cream the fat and sugar till gets light, creamy white and fluffy. Take curd, water and salt make solution and add it in the creamed fat. Add dextrose solution and then sieved flour. Knead the mixture very well. Then shape into smooth round ball, place it on greased baking sheets or baking tray and bake at 180 - 190°C for about 10 – 15 min. to get good colour then remove from the oven, cool it and pack in plastic bags for storage. Such types of nankhatai can remain in good condition upto 10 – 15 days. The nutritive value of such type of cookies and biscuits are given in **Table 60**.

Sorghum melting moments

Ingredient	Quantity
Malted sorghum flour	55 g
Maida	140 g
Sugar powder	85 g
Butter	115 g
Baking powder	1.5 tea spoon
Egg	1 No
Cashew nut	15 g
Vanilla essence	Few drops
Ammonium bicarbonate	1 g
Sodium bicarbonate	2 g

Procedure for the preparation of sorghum melting moments

Sieve sorghum flour, maida, baking powder, ammonium bicarbonate, and sodium bicarbonate together. Then cream the fat and sugar till gets light, creamy white and fluffy. Add slightly beaten egg and vanilla essence and beat well. Take flour and mix with above mixture and make smooth dough. Divide the dough into small balls and roll in crushed cashew nut and put in a greased and floured baking sheet or tray. Bake in a oven at 180 - 190ºC for 15 – 20 min, cool and pack in LDPE bags. The nutritive value of such type of cookies and biscuits are given in **Table 60**.

Sorghum peanut cookies

Ingredient	Quantity
Malted or plain sorghum flour	150 g
Maida	350 g
Sugar powder	250 g
Peanut powder	80 g
Butter	200 g
Salt	5 g
Baking powder	1.5 tea spoon
Egg or milk powder	1 No or 10 g

Procedure for the preparation of sorghum peanut cookies

Sieve sorghum flour, maida and baking powder together. Then cream the fat and sugar till gets light, creamy white and fluffy. Add slightly beaten egg and vanilla essence and beat well. Take flour, peanut powder and mix with above mixture and make smooth dough. Add milk if required. Divide the dough into small balls and put in a greased baking sheet or tray. Bake in oven at 180 - 190°C for 8 – 10 min, cool and pack in LDPE bags. The nutritive value of such type of cookies and biscuits are given in **Table 60**.

Sorghum masala biscuits

Ingredient	Quantity
Malted or plain sorghum flour	300 g
Maida	300 g
Sugar powder	100 g
Fat	275 g
Green chilli	20 g
Salt	5 g
Curd	150 g
Ammonium bicarbonate	1 g
Sodium bicarbonate	5 g
Ajwan and cumin seed powder	5 g
Coriander leaves	10 g
Dextrose liquid	50 ml

Procedure for the preparation of sorghum masala biscuits

Sieve sorghum flour, maida, chilli powder, anjawan, cumin powder, ammonium carbonate, and sodium carbonate together Then cream the fat and sugar till gets light, creamy white and fluffy. Add curd, dextrose liquid, coriander paste and mix well. Take flour and mix with above mixture and make smooth dough. Add milk if required. Divide the dough into small balls and put in a greased baking sheet or tray. Bake in oven at 180°C for10– 15 min, cool and pack in LDPE bags. The nutritive value of such type of cookies and biscuits are given in **Table 60**.

Sorghum Salt biscuits

Ingredient	Quantity
Malted or plain sorghum flour	400 g
Wheat maida	100 g
Sugar powder	150 g

Ingredient	Quantity
Fat	200 g
Eggs	2
Salt	10 g
Baking powder	½ tea spoon
Ammonium bicarbonate	4 g
Water	As required

Procedure for the preparation of sorghum salt biscuits

First sieve sorghum flour, baking powder and maida together. Mix the all flours with fat. Dissolve ammonia bicarbonate, salt, essence and sugar powder in small quantity of water. Add this material to the mixture of flour and fat and make smooth dough with addition of emulsified eggs. Make this dough as a flat 1/8 inch width, and then cut into different shapes using biscuit cutters. Place these biscuits on a greased tray and bake at 170 – 180°C temperature for 15 – 20 min in an oven. Cool the salty biscuits and store at room temperature in plastic bags and use when required. Such types of biscuits can remain in good condition for 15 days at room temperature. The nutritive value of sorghum salt biscuits is given in **Table 60**.

Sorghum Sweet biscuits

Ingredient	Quantity
Malted or plain sorghum flour	300 g
Wheat maida	200 g
Sugar powder	300 g
Fat	300 g
Salt	To taste
Baking powder	1 tea spoon
Ammonium bicarbonate	2 g
Essence	Few drops
Water	If required

Procedure for the preparation of sorghum sweet biscuits

First sieve sorghum flour, baking powder, salt, ammonium bicarbonate and maida together. Rub the fat thoroughly and add sugar powder and rub well. Then add few drops of essence and add flour slowly and mix well to a fine dough. Make this dough as a flat 1/8-inch width, and then cut into different shapes using biscuit cutters. Place these biscuits on a greased tray and bake at 170 – 180°C temperature for 15 – 20 min in an oven. Cool the sweet biscuits and store at room temperature in plastic bags and use when required. Such types of biscuits can remain in good condition for 15 days at room temperature. The nutritive value of sorghum sweet biscuits is given in **Table 60**.

Sorghum Coconut biscuits

Ingredient	Quantity
Malted or plain sorghum flour	500 g
Maida	100 g
Desiccated coconut	200 g
Sugar powder	300 g
Fat	300 g
Salt	To taste
Baking powder	2 tea spoon
Ammonium bicarbonate	4 g
Essence	Few drops
Water	If required

Procedure for the preparation of sorghum coconut biscuits

First sieve sorghum flour, baking powder, ammonium bicarbonate, salt and maida together. Rub the fat thoroughly and add sugar powder and rub well. Then add few drops of essence and add flour slowly with coconut flour and mix well to a fine dough. Make this dough as a flat 1/8-inch width, and then cut into different shapes using biscuit cutters. Place these biscuits on a greased tray and bake at 170 – 180°C temperature for 15 – 20 min in an oven. Cool the sweet biscuits and store at room temperature in plastic bags and use when required. Such types of biscuits can remain in good condition for 15 days at room temperature. The nutritive value of sorghum coconut biscuits is given in **Table 60**.

Sorghum Peanut biscuits

Ingredient	Quantity
Malted or plain sorghum flour	500 g
Maida	100 g
Peanuts	200 g
Sugar powder	300 g

Ingredient	Quantity
Fat	300 g
Salt	To taste
Baking powder	2 tea spoon
Ammonium bicarbonate	4 g
Essence	Few drops
Water	If required

Procedure for the preparation of sorghum peanut biscuits

First sieve sorghum flour, baking powder, ammonium bicarbonate, salt and maida together. Cut the peanuts into small pieces and mix with flour. Rub the fat thoroughly and add sugar powder and rub well. Then add few drops of essence and add flour slowly with coconut flour and mix well to a fine dough. Make this dough as a flat 1/8 inch width, and then cut into different shapes using biscuit cutters. Place these biscuits on a greased tray and bake at 170–180ºC temperature for 15–20 min in an oven. Cool the sweet biscuits and store at room temperature in plastic bags and use when required. Such types of biscuits can remain in good condition for 15 days at room temperature. The nutritive value of sorghum peanut biscuits is given in **Table 60**.

The general procedure for the preparation of biscuits and cookies is given in **Figure 16**.

Sorghum *shankarpale*

Ingredient	Quantity
Sorghum flour	150 g
Maida	300 g
Sugar powder	100 g
Fat	100 g
Salt	3 g
Ammonium bicarbonate	2 g
Baking powder	¼ tea spoon
Cardamom powder	5 g
Dextrose liquid	75 g
Water	35 ml

Sorghum grains + other grains

↓

Cleaning

↓

Processing (malting)

↓

Grind into fine flour

↓

Sieve if necessary

↓

Take course/fine flour for formulations

Flours (100%)	Colour (0.01%)
Grind sugar (35%)	Essence (0.1%)
Fat (20%)	Thickening agent
Chemical leavening (0.2%)	Emulsifying, Antioxidants

↓

Water + Milk (30%)

↓

Mix into smooth dough and rest for 30 min.

↓

Roll into sheet (0.5 cm)

↓

Cut into round biscuits or cookies

↓

Bake at 175 – 190 C for 10 – 15 min in preheated oven (till it get brown bake colour)

↓

Cool and pack in airtight container or in poly pouch after vacuumization

↓

Storage

↓

Marketing

Figure 16: General flow sheet for preparation of sorghum biscuits/cookies

Procedure for the preparation of sorghum *shankarpale*

Sieve sorghum flour, maida and baking powder together. Then cream the fat and sugar till gets light, creamy white and fluffy. Add water and cardamom powder. Then mix sieved flour and make smooth dough. Roll the dough into 0.5 cm thickness and cut with knife and make small squares shapes. Fry in oil/fat at 170ºC for10– 15 min, cool and pack in LDPE bags. The nutritive value of such type of *shankarpale* is given in Table 60.

Sorghum sponge cake

Ingredient	Quantity
Malted or plain sorghum flour	40 g
Maida	75 g
Sugar powder	115 g
Fat	75 g
Salt	1 g
Egg	2 No
Milk powder	20 g
Baking powder	¼ tea spoon
Vanilla essence	Few drops
Water	As required

Procedure for the preparation of sorghum sponge cake

Sieve sorghum flour, maida, milk powder and baking powder together. Then cream the fat and sugar till gets light, creamy white and fluffy. Beat the eggs and beat into cream fat and pour into the sieved flour, add few drops of essence. Add water if required to get a dropping consistency. Pour the mixture into a waxed paper lined cake tin and bake at 175°C for10 – 12 min, cool and pack in LDPE bags. The nutritive value of such type of sponge cake is given in **Table 60**.

Sorghum Fruit cake

Ingredient	Quantity
Malted or plain sorghum flour	150 g
Maida	50 g
Pulpy fruits (pulp)	100 g
Sugar powder	70 g

Ingredient	Quantity
Fat	75 g
Salt	1 g
Egg	3 No
Milk powder	10 g
Baking powder	¼ tea spoon
Sodium bicarbonate	1 tea spoon
Vanilla essence	Few drops
Milk	As required

Procedure for the preparation of sorghum fruit cake

Sieve sorghum flour, maida, milk powder and baking powder together. Then cream the fat and sugar till gets light, creamy white and fluffy. Beat the eggs and beat into cream fat and pour into the sieved flour, add few drops of essence and pulp of the fruit. Add milk if required to get a dropping consistency. Pour the mixture into a waxed paper lined cake tin and bake at 175°C for10 – 12 min, cool and pack in LDPE bags. The nutritive value of such type of fruitcake is given in **Table 60**.

INDUSTRIAL/NUTRACEUTICAL PRODUCTS FROM SORGHUM

Malting of sorghum for brewing industry

The malting of sorghum reduced the content of starch and increased the reducing sugars. The malting also improved *in vitro* protein digestibility (IVPD) from 80.30 to 83.60%. The malted cereals and pulses are mostly utilized in the preparation of weaning food (infant food). Malted cereals as well as pulses make all major nutrients easily digestible by hydrolysis process and convert into very simplest form, which can absorb in the digestible system very easily **(Table 69)**.

Table 69: Proximate composition of sorghum and sorghum malt.

Constituent (%)	M 35-1		Swati		CSH 5		PVK-801	
	Grain	Malt	Grain	Malt	Grain	Malt	Grain	Malt
Moisture	10.13	8.13	11.02	8.48	10.83	8.93	10.00	7.91
Crude protein	11.60	11.60	10.70	10.70	11.30	11.30	9.45	9.42
Crude fat	2.93	2.87	2.83	2.78	2.98	2.81	3.00	2.90
Crude fibre	2.13	2.10	2.35	2.10	2.48	2.39	2.10	2.00

Constituent (%)	M 35-1		Swati		CSH 5		PVK-801	
	Grain	**Malt**	**Grain**	**Malt**	**Grain**	**Malt**	**Grain**	**Malt**
Starch	76.23	68.13	77.24	67.83	76.81	66.35	74.20	59.50
Reducing sugar	0.20	1.40	0.10	0.70	0.10	1.40	0.30	0.70
Total minerals	1.70	1.40	1.61	1.43	1.30	1.10	1.20	0.90
Malting losses	-	13.67	-	16.15	-	17.50	-	19.00

Sorghum grains

↓

Soaking

↓

Germination

↓

Drying of malted grains

↓

Ground to fine flour

↓

Malted flour

↓

Add cold water to prepare slurry

↓

Add more pounded/steeped sorghum grains in it

↓

Keep for fermentation

↓

Filtration

↓

Bottling of malt beer

↓

Storage

Figure 17. Flowchart for preparation of sorghum beer.

Use of sorghum in brewing industry (sorghum beer)

In sole sorghum beer (without adjunct), the bitterness and free amino nitrogen content were low compared to sorghum adjunct beer, which had highest free amino nitrogen content and essential for the beer flavour **(Table 70).**

Brewing is said to raise the nutritional value of sorghum since it adds Vitamins, neutralizes most of the tannins, hydrolyzes the starch to more digestible forms and increases the availability of minerals and vitamins. South African studies indicate that iron is 12 times more available in sorghum beer than in boiled sorghum gruel, riboflavin may be almost twice and thiamine almost a third more available while niacin's availability remains unchanged **(Figure 17).**

Enzymes in the malt begin to saccharify the starch. Bacteria then act on a part of the sugar to form lactic acid and some B vitamins. This sweet wort may be drunk after less than one day's fermentation. It contains more B vitamins and negligible amount alcohol. After the first day the mixture is heated and various fungal spores germinate and lead to the production of hydrolytic enzymes. A little millet or sorghum may be specially germinated, then dried as a yeast source. The yeasts multiply and fermentation develops, with the production of alcohol, more B vitamins and new proteins, while the bacteria continue to develop lactic acid. The brew may be drunk after 4-5 days, and in some areas more malted grain may be added to it a day before it is used.

Table 70: Chemical composition of beer from sorghum

Constituent	Malt beer	Adjunct beer
Protein (%)	2.68	1.98
Free amino nitrogen (%)	74.61	119.01
Total sugars (%)	1.54	0.95
Reducing sugars (%)	0.05	0.02
Alcohol (%)	4.66	4.50
Colour (EBC)	5.89	8.58
Specific gravity	1.03	0.98
Bitterness (BU)	6.56	7.58

Standardization of suitable process for the extraction and recovery of starch from moulded and blackened sorghum grains

The black sorghum grains were soaked in distilled water for 24 hrs to 168 hrs separately at ambient temperature (27 ± 2°C) and 60°C. The soaking water was changed after every 24 hrs. The soaking grains were blended with water (1:2, w/v) in a laboratory blended for 2 min. The blending was repeated for two more minutes after a pause of 1 min. The slurry was screened through a double-layered muslin cloth. The residue on the muslin cloth was washed with additional water to solubilize and filterate out the remaining starch. The filterate was centrifuged at 8000 x g for 30 min. The supernatants were discarded and the residue was suspended in 0.2 percent sodium hydroxide solution to solubilize the proteins.

The contents were allowed to stand for 3 to 4 hrs to settle the starch. The yellowish upper sodium hydroxide layer was decanted and the settled starch was resuspended in sodium hydroxide until the sodium hydroxide layer became colourless. The settled starch was then washed with water and the contents were screened through the 0.045 mm sieve. The filterate was centrifuged and the starch obtained was recorded as residue in centrifuge tubes and dried at ambient temperature (**Figure 18**).

A maximum absorption of water was observed in the grins soaked at 60ºC for 96 hrs. However, the highest starch yield was obtained when the grains were soaked at 60ºC for 144 hrs. The dilute alkali (0.2%) was found to be superior grain soaking and residue washing medium than the dilute acids or water. A significant higher starch yield was obtained with the use of 0.045 mm sieve (59.3%) as a slurry filtration device as compared to that of muslin cloth (33.8%).

The starch obtained with muslin cloth exhibited slightly higher purity but significantly lower recovery. A soaking of grains for 144 hrs at 60^0C in 0.2 percent NaOH followed by grinding in water and filtration of the slurry through 0.045 mm sieve, centrifugation of the filterate at 8000 x g for 30 min, washing of the residue in 0.2 percent NaOH followed by water washing, centrifugation and drying of the residue has been standardized to obtain a maximum of 62.2 percent bright starch from molded and blackened sorghum. The product contained 91.6 percent starch.

A process to get white starch from the mould-infected sorghum with black or red / brown discoloration is described. It involves soaking of grains in 0.2% solution of H_2SO_4 or NaOH (1:2, w/v) for 12 hrs at 60ºC followed by blending in the same extract for 2-3 min and filtration of slurry through 0.045 mm sieve (**Figure 19**). The extracted starch is recovered by centrifugation of the filterate at 8000 x g for 20 min, washing the starch-pellet with water to remove residual acid/alkali and drying the product at 40ºC. Sodium hydroxide was suitable for extraction of starch from grains with black and red/brown discoloration while H_2SO_4 was suitable for only blackened grains. The physico-chemical properties of sorghum starch are given in **Table 71 to 78.**

Table 71: Water absorption by mould infected blackened sorghum.

Grain soaking temperature ºC	Weight gain after soaking of grains (hrs)			
	6	12	24	Mean
30	24.8 + 0.6	31.9 + 0.7	38.7 + 0.8	31.8
40	27.2 + 0.7	33.6 + 0.6	40.1 + 0.4	33.7
50	37.7 + 0.6	36.4 + 0.8	42.6 + 0.7	38.9
60	39.9 + 0.7	42.9 + 0.6	43.0 + 0.4	41.9
70	41.1 + 0.4	43.0 + 0.5	43.1 + 0.6	42.4
Mean	33.1	37.6	41.5	37.7

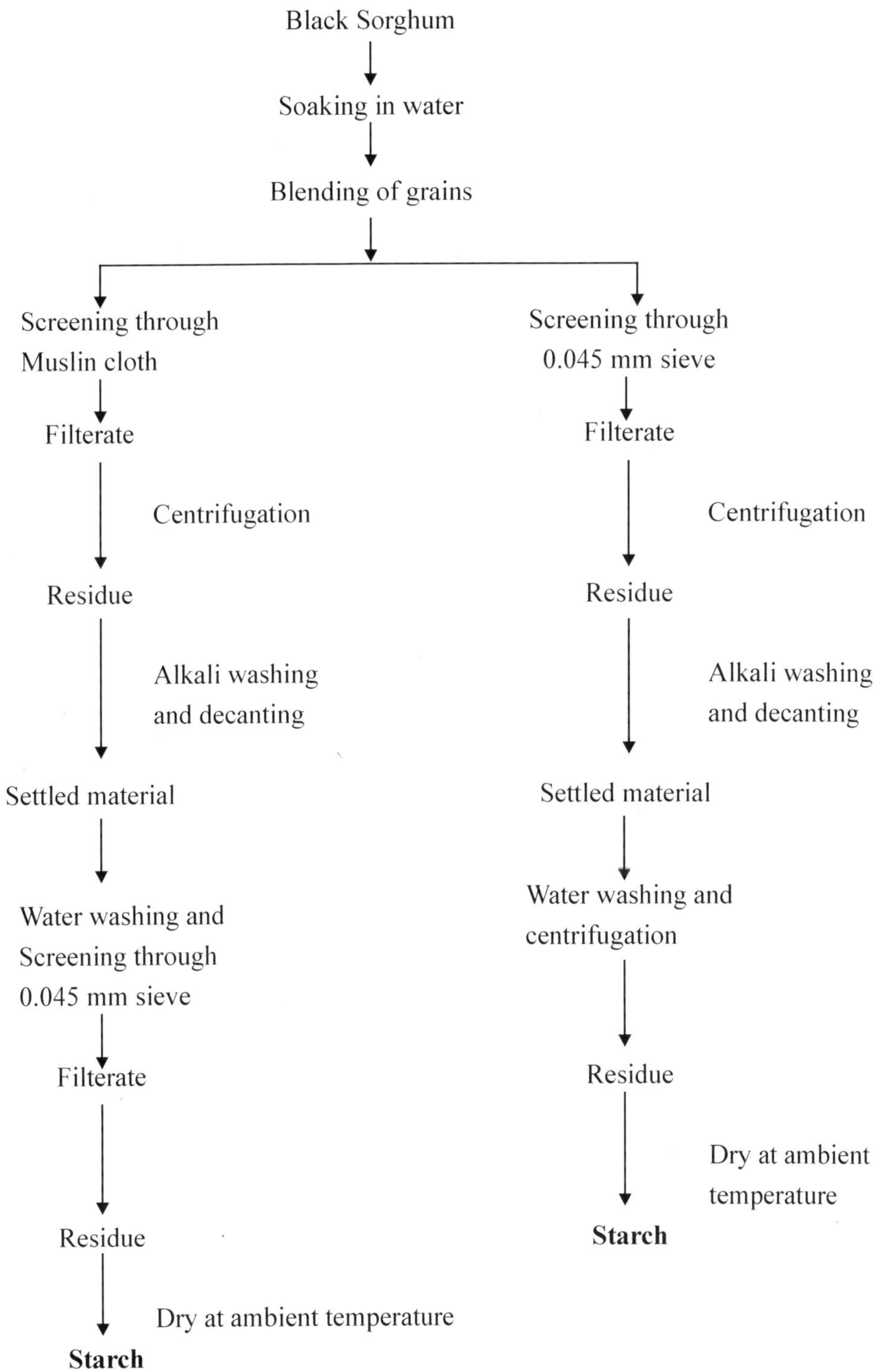

Figure 18: Standardization of starch extraction processes from sorghum.

Table 72: Effect of soaking medium on extraction and recovery of starch from black sorghum.

Grain soaking medium (%)	Extracted starch (%)
NaOH, 0.1	52.00 ± 2.4
H_2SO_4, 0.1	23.38 ± 1.8
Lactic acid, 0.1	34.4 ± 1.9
Water	33.8 ± 2.1

Table 73: Effect of different soaking periods and extraction media on yield of starch from black sorghum.

Soaking and extraction medium	Starch yield (%) after soaking grains (hrs)		
	6	12	24
Water	45.7 ± 1.6	52.2 ± 1.8	53.0 ± 1.7
Lactic acid, 0.2%	49.4 ± 1.9	56.7 ± 1.5	57.2 ± 1.5
Potassium metabisulphite, 0.2%	50.0 ± 1.8	59.2 ± 1.6	60.0 ± 1.7
Sodium hydroxide, 0.2%	52.8 ± 1.7	64.0 ± 1.4	64.3 ± 1.6
Sulphuric acid, 0.2%	53.7 ± 1.9	63.8 ± 1.8	64.1 ± 1.5

Table 74: Effect of increasing concentration of alkali as soaking medium on starch extraction and recovery.

NaOH concentration (%)	Starch recovery (%)
0.1	52.0 ± 2.4
0.2	54.4 ± 2.6
0.5	54.8 ± 2.8

Table 75: Effect of screening method on recovery of starch.

Screening method	Starch recovery (%)
Muslin cloth	33.8 ± 1.6
Sieve (0.045 mm)	59.3 ± 2.7

Table 76: Effect of period and temperature of soaking of grains on the extraction and recovery of starch from black sorghum grains.

Soaking period (hrs)	Starch recovery (%) at	
	27 ± 2°C	60°C
24	17.9 ± 0.50	18.4 ± 0.90
48	20.2 ± 1.30	20.8 ± 1.10
72	23.7 ± 0.80	24.7 ± 0.80
96	27.2 ± 1.60	27.6 ± 1.20
120	28.5 ± 0.80	32.1 ± 1.60
144	32.0 ± 0.95	33.8 ± 0.85
168	32.4 ± 0.80	34.0 ± 0.60

Table 77: The yield, content and recovery of starch by different methods of extraction.

Sr. No	Type of method used			Percent starch in black grain	Starch product from grain, %	Actual starch content in extracted product, %	Starch recovery on initial grain basis, %
	SM	BM	RW				
1.	W	W	N	64.7	33.8 ± 1.20	96.0 ± 1.59	50.2 ± 1.60
2.	N	W	N	64.7	52.0 ± 1.80	93.8 ± 1.65	75.4 ± 1.75
3.	N	N	W	64.7	48.3 ± 1.90	94.7 ± 1.57	70.7 + 1.70
4.	W	W	N	64.7	59.3 ± 1.80	92.0 ± 1.83	84.3 ± 1.81
5.	N	W	N	64.7	62.2 ± 1.95	91.6 ± 1.92	88.0 ± 1.79
6.	N	N	W	64.7	61.8 ± 1.10	92.2 ± 1.50	88.1 ± 1.67

Method 1 to 3 with muslin cloth, Method 4 to 6 with 0.045 mm sieve.
SM = Soaking medium, BM = Blending medium, RW = Residue washing medium
W = Water, N = NaOH (0.2%).

Black/Mould infected discoloured sorghum grains

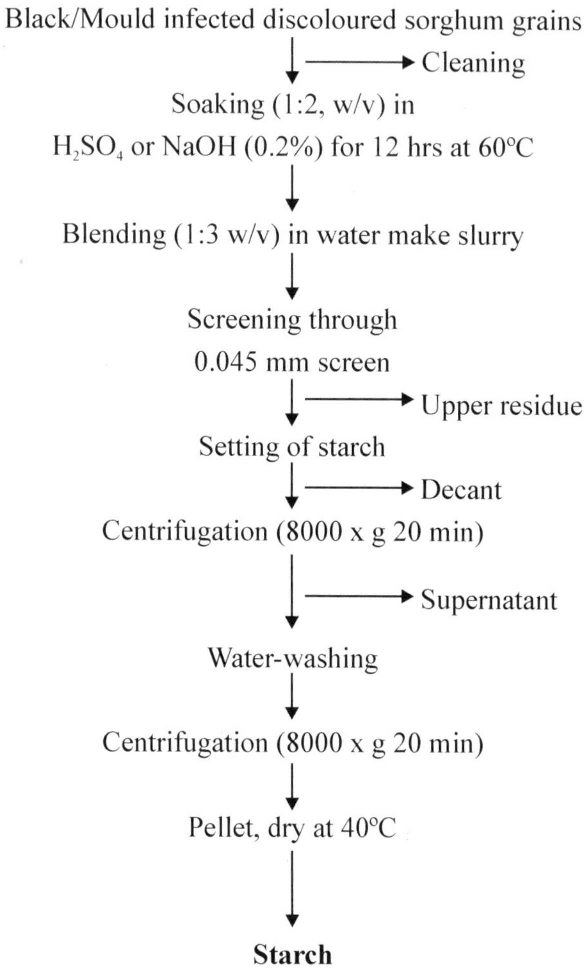

Figure 19: Standardized process for extraction of starch from black/mould-infected discoloured sorghum grains.

Table 78: Chemical composition of starch extracted by different method.

Starch extraction method	Starch, %	Moisture, %	Protein, %	Total sugars, %	Total ash, %
1.	96.0 ± 1.59	4.5 ± 0.33	0.47 ± 0.10	0.47 ± 0.02	0.50 ± 0.03
2.	93.8 ± 1.65	4.0 ± 0.35	1.03 ± 0.09	0.45 ± 0.01	0.25 ± 0.02
3.	94.7 ± 1.60	5.0 ± 0.29	1.24 ± 0.08	0.43 ± 0.03	0.10 ± 0.04
4.	92.0 ± 1.57	4.5 ± 0.25	1.06 ± 0.06	0.41 ± 0.05	0.39 ± 0.02

Starch extraction method	Starch, %	Moisture, %	Protein, %	Total sugars, %	Total ash, %
5.	91.6 ± 1.92	4.8 ± 0.30	1.23 ± 0.07	0.45 ± 0.04	0.40 ± 0.03
6.	92.2 ± 1.50	4.8 ± 0.25	1.66 ± 0.06	0.43 ± 0.03	0.42 ± 0.01

Method 1 to 3 with muslin cloth, Method 4 to 6 with 0.045 mm sieve.
Methods: 1. WWN, 2. NWN, 3. NNW, 4.WWN, 5. NWN, 6. NNW
SM = Soaking medium, BM = Blending medium, RW = Residue washing medium
W = Water, N = NaOH (0.2%).

Table 79: Functional properties of extracted starch from black sorghum.

Starch extraction method	Water absorption capacity (ml/g)	Oil absorption capacity (ml/g)	Gelatinization temperature (^{0}C)
1.	0.90 ± 0.06	0.99 ± 0.05	78.6 ± 1.2
2.	1.12 ± 0.07	1.18 ± 0.06	77.2 ± 1.1
3.	1.20 ± 0.06	1.11 ± 0.07	78.0 ± 1.01
4.	0.92 ± 0.07	1.32 ± 0.04	80.2 ± 1.10
5.	0.93 ± 0.05	0.99 ± 0.06	76.8 ± 1.30
6.	0.95 ± 0.06	1.01 ± 0.05	78.3 ± 1.15

Method 1 to 3 with muslin cloth, Method 4 to 6 with 0.045 mm sieve.
Methods: 1. WWN, 2. NWN, 3. NNW, 4.WWN, 5. NWN, 6. NNW
SM = Soaking medium, BM = Blending medium, RW = Residue washing medium
W = Water, N = NaOH (0.2%).

Table 80: Physicochemical properties of starch extracted from black or red/ brown discoloured sorghum.

Parameter	Whole grains	H_2SO_4 extracted starch	NaOH extracted starch
Moisture, %	9.5	3.4	4.0
Starch, %	67.6	94.5	94.0
Protein, %	9.5	0.7	0.8
Ash, %	1.7	0.3	0.4
Fat. %	2.4	0.9	0.9
Fiber, %	2.2	-	-
GT, °C	-	77.5	78.3
Colour	Black	White	White
Appearance	Red/brown	Pinkish	White

*GT = Gelatinization temperature.

Starch and maltodextrin from sorghum

The starch prepared from black sorghum grain have a vital role and potential raw material for the production of maltodextrin and it may be explored for industrial or commercial exploitation in different food products as fat replacer to reduce the calorific value. Metabolisms of maltodextrins are similar to starch. Food products for diabetic patients can be prepared by using maltodextrins as a fat replacer. Maltodextrins gives hypoglycemic activity and it also a promising food substance to heart patients because it acts as a very good hypocholesterlemic agent.

The maximum amount of maltodextrin was liberated at 4.5% concentration of hydrochloric acid hydrolysis process at 120 minutes interval i.e., 95 per cent and 92 per cent for PVK 801 and CSH 9 sorghum varieties and their dextrose equivalents were 9 and 8 respectively. The maltodextrin contained 98 and 97 per cent total carbohydrate in PVK 801 and CSH 9 respectively **(Table 81).** The results of reheological attributes particularly the pasting behaviour of maltodextrin revealed that the viscosity at 40°C was found to be maximum and then decreased as the temperature increased. The microscopic examination of maltodextrin granules revealed that they are mostly irregular shaped particle aggregates with 4 – 5 μ in size (**Table 82).** The results indicated that good quality cookies and toffees could be prepared using maltodextrin with 3 per cent replacement of shortening. Maltodextrins are metabolized similar to starch. Thus, fat replacers like maltodextrin can be recommended for use in food products for diabetics, it exhibits hypoglycemic activity. Moreover, it will be a very promising for heart patients also as it acts as a very good hypocholesterlemic agent. The comparative characteristics of starch and maltodextrin of sorghum is given in **Table 83 and Figure 20**.

Table 81: Chemical composition and physical parameters of sorghum starch.

Parameter	Content (%)	
	PVK-801	**CSH-9**
Moisture	9.80	9.50
Protein	1.10	1.29
Fat	0.90	0.95
Ash	0.24	0.45
Crude fiber	0.73	0.67
Total carbohydrate	96.81	95.60
Yield of starch	75.50	72.85
Granule shape	Polygonal to round	Polygonal to round
Granule size	8.0-9.5 μ	8.0-9.1 μ
Colour of starch	Bright white	Mild white
Iodine affinity	4.4	3.9

Parameter	Content (%)	
	PVK-801	CSH-9
Bulk density	0.59	0.58
Molecular weight		
Amylose (daltan)	2,45,000 ± 10,000	2,45,000 ± 10,000
Amylopectin (daltan)	> 0.59 x 10^6	> 0.59 x 10^6

Table 82: Physico-chemical constituents of maltodextrin.

Parameter	Content (%)	
	PVK-801	CSH-9
Moisture	8.0	7.6
Protein	0.05	0.07
Carbohydrate	98.0	97.0
Fat	0.07	0.08
Ash	0.6	0.7
Bulk density	0.579	0.575
Granule shape	Irregular	Irregular
Granule size	3-5 μ	4-5 μ
Granule colour	Bright white	White
Molecular weight		
Amylose (daltan)	1,75,000 ± 5000	1,75,000 ± 5000
Amylopectin (daltan)	> 0.59 x 10^6	> 0.59 x 10^6
Dextrose equivalent	9	8

Table 83: Physico-chemical properties of sorghum starch and maltodextrin.

Constituent	Starch	Maltodextrin
Moisture (%)	9.50-9.80	7.5-8.0
Protein (%)	1.10-1.29	0.05-0.07
Fat (%)	0.90-0.95	0.07-0.08
Ash (%)	0.24-0.45	0.6-0.7
Fiber (%)	0.67-0.73	0.4-0.5
Total carbohydrates (%)	95.60-96.81	97-98
Yield (%)	73-76	-
Granule shape	Round, polygonal	Irregular
Granule size (μ)	8.0-9.5	3.0-5.0
Colour of granule	Bright white	Mild white

Constituent	Starch	Maltodextrin
Iodine affinity	3.9-4.4	-
Bulk density	0.58-0.59	0.57-0.59
Molecular weight of Amylose (daltan)	2,45,000	1,75,000
Molecular weight of Amylopectin (daltan)	$> 0.59 \times 10^6$	$> 0.59 \times 10^6$

Sorghum grains

↓

Cleaning (Removal of foreign matters)

↓

Steeping in water (20 hrs, at 50°C and 4.5 pH)
(750 ppm KMS, 0.002 M $HgCl_2$, few drops of toluene in water)

↓

Washing with clean water

↓

Blending / preparation of slurry (1:3 w/v, seed : water)

↓

Separation of starch
(Squeezing of slurry through muslin cloth and then passing through 150 μ sieve)

↓

Protein denaturation
(Add 1.5 g NaCl and 5 ml Toluene in slurry)

↓

Shaking for 1 hr

↓

Centrifugation at 4200 RPM for 15 min.

↓

Isolation of starch

↓

Drying (Vacuum drying)

↓

Air drying for over night

↓

Starch

↓

Acid hydrolysis
(4.5 % HCl conc. 50°C temp. and 120 min. time)

↓

Isolation → (Vacuum drying) → Maltodextrin

Figure 20: Flow diagram for the isolation and purification of starch and maltodextrin.

Production of high fructose syrup from sorghum starch

The sorghum genotypes cultivated in *kharif* season contain about 68 to 76 per cent starch. The starch is a potential raw material being used in several cloth, paper, dye as well as in food industry in India and foreign country also. Therefore, the production of starch from *kharif* sorghum is one of the most prominent sources because most of the high yielding hybrids mostly cultivated during *kharif* season. For the preparation of value addition products from rain damaged *kharif* season sorghum and also to exploit it for industrial utilization, preparation of starch and its conversion into other valuable products such as high fructose syrup, maltodextrin, sweeteners, carboxymethylated starch, glucose, and sorbitol is most essential.

Fructose is the sweetest sugar among the natural sugars. The relative sweetness of fructose is a function of concentration, temperature and pH, with the maximum sweetness value being found in cooled, dilute solutions that is neutral or slightly acidic. Fructose has a short sweetness duration that tends to enhance the aroma and flavours of fruits. The high fructose syrup can be prepared from sorghum **(Figure 21a and 21b)** as well as sweetener too **(Figure 21c)**. The refractive index for high fructose syrup (HFS) prepared from CSH 18 starch (1.453) was same as that of HFS from glucose sample (1.455). Colour of the syrup prepared from starch was yellowish and the colour of the HFS prepared from control glucose sample was whitish yellow **(Table 84).**

Commercially sucrose is usually marketed as sucrose or medium invert sugar, which is liquid and contains 50 parts sucrose, 25 parts dextrose and 25 parts fructose. High fructose sorghum syrup is similar to total sucrose invert; therefore, it replaced sucrose on weight-to-weight basis. Due to the refining to remove non-sweetener components and because of excessive microbiological control, HFS is a very high quality liquid sweetener. The predominant user of HFS is the soft drink industry. The mango squash prepared by using HFSS had recorded higher preference for all the attributes than that of control. The high fructose sorghum syrup can be effectively used to replace sucrose in the products like fruit squashes.

Sorghum starch

↓ Dextrinization with acid hydrolysis
Liquification by -amylase
Saccharification with Glucoamylase

Glucose syrup (95-96%) Dextrose

↓ Refining
Demineralization
Decolourisation

Purification

↓

Purified glucose and syrup

↓

Isomerization

↓ With glucose isomerase
Temperature 60°C
pH 6.8-7.0

High fructose sorghum syrup (HFSS)

↓

Purification

↓ Decolourization using
carbon celite column

HFSS (45 % D.S.)

↓

Evaporation

↓

HFSS (54-55 % D.S.)

Figure 21a: Production of high fructose syrup from sorghum starch

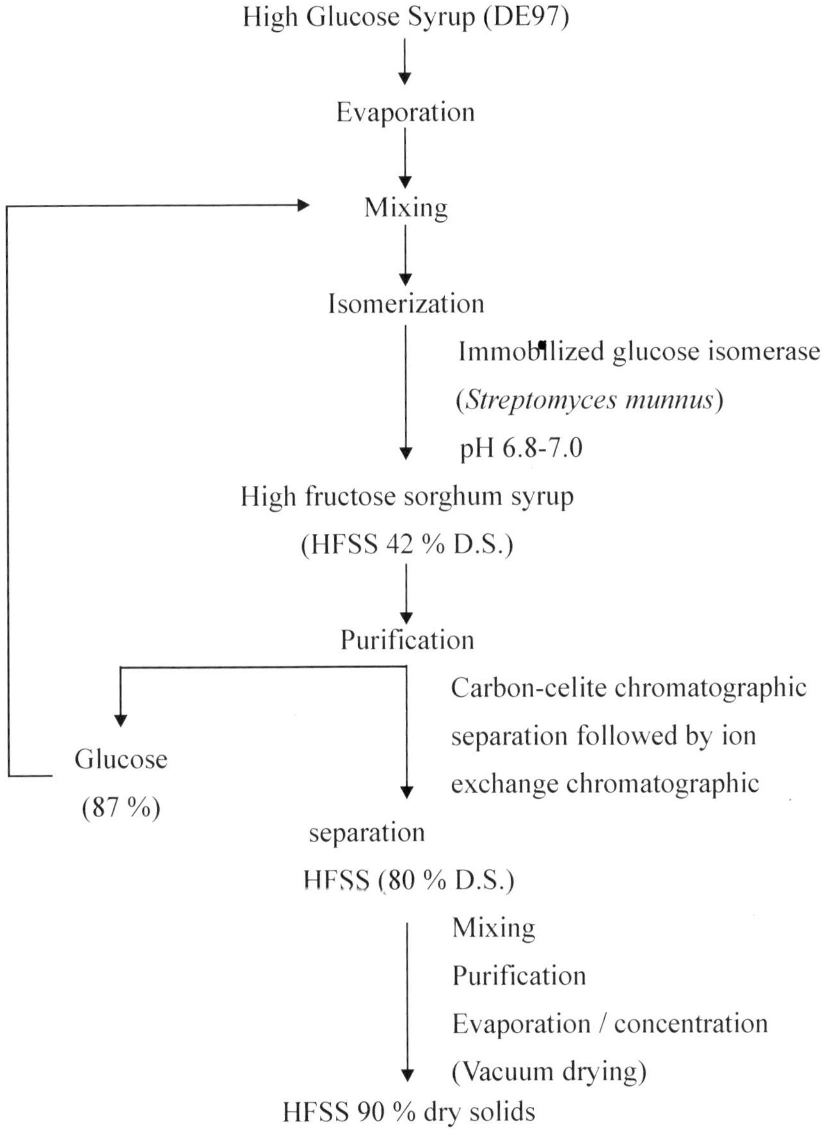

Figure 21b: Production of high fructose syrup using immobilized glucoisomerase (*Streptomyces munnus*) at optimum experimental conditions

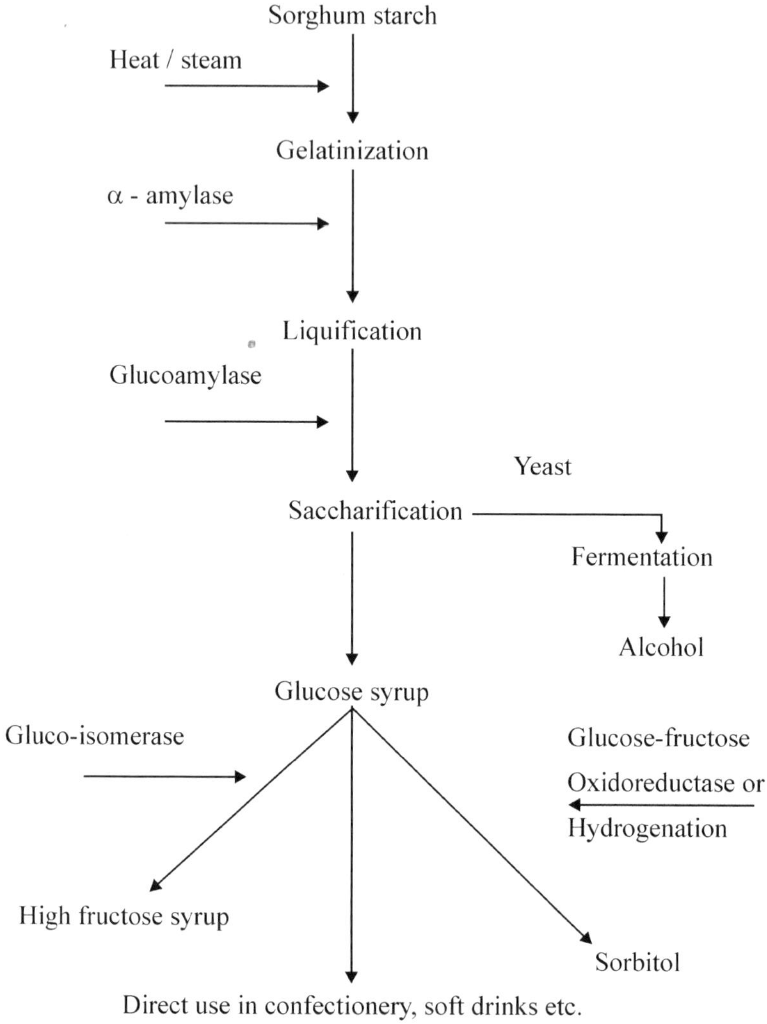

Figure 21c: Sweetener production from starch

Table 84: Physico-chemical properties of high fructose syrup.

Parameter	HFSS from CSH-18	HFS control (glucose)
Refractive index	1.453	1.455
Colour	Yellowish	Whitish yellow
Dry solids (%)	67.93	70.00
Moisture (%)	32.06	30.00
Ash (%)	0.08	0.05
Acidity (%)	0.11	0.09
°Brix (TSS,%)	58.33	60.00

Production of sorbitol from sorghum starch

Sorbitol solution appears to be a clear, colourless and syrup liquid with pleasantly sweet taste. The solution was essentially odourless. The sorbitol content of the solution was found to be 66 per cent **(Table 85)**. Study of the composition of the prepared solution revealed that the solution contents 0.21 per cent reducing sugars and 0.70 per cent total sugars. In order to assess the importance and feasibility in food and other industries, the viscosity profile of sorbitol solution is considered as one of the important parameters. Sorbitol being a monosaccharide polyol, contributes little to the viscosity of the solution to which it is added. The viscosity of sorbitol solution over a wide range of shear rates and temperature has been carried out by using Haake's Rotoviscometer (RV 20). The flow behaviour of sorbitol solution at various temperatures (20 – 26°C) was investigated. The results indicated that the viscosity of the sorbitol solution linearly decreases with the increase in the temperature and shear rate **(Table 86)**.

Various functional properties of sorbitol were extensively studied. The refractive index of the solution (at 20°C) was found to be 1.45. The relative sweetness of the 10 per cent solution was compared with the sweetness of standard sucrose. The relative sweetness of the solution was found to be 60:100. The data on the cooling effect of the solution revealed that the solution gives cooling effect within six seconds of putting it in mouth. The solution was found to be thermally stable up to 160 – 180°C **(Table 87)**. The flash point of the sorbitol solution was found to be 300°C and the specific gravity of the solution was 1.28 at 25°C. The non-carcinogenic properties of sorbitol are making it increasing accepted sweetener in the insulin-independent mechanism of sorbitol make it an ideal sweetener for the diabetes. Sorbitol produced from sorghum starch hydrolyzes by chemical hydrogenation in presence of nickel catalyst and was used as sweetener. The milk chocolate prepared by using sucrose had scored higher value for all the quality attributes than the milk chocolates prepared by using sorbitol and sucrose in combination. Milk chocolate manufactured from sorbitol resulted in low-viscosity chocolate base, which was corrected by using additives (starch or guar gum 2 per cent).

Table 85: Biochemical properties of sorbitol solution

Parameter	Observations
Appearance	Clear, colourless, syrup liquid
Colour	Watery white
Taste	Pleasantly sweet
Odour	Essentially odourless
Assay (%) D-sorbitol	66.0
Reducing sugars (%)	0.21

Parameter	Observations
Total sugars (%)	0.70
Residues on ignition (%)	0.10
Water (%)	29.5
pH (10% solution)	6.5

Table 86: Viscosity profile for sorbitol syrup.

Temperature (°C)	Apparent viscosity (Cps)			
	Shear rate (S^1)			
10	4.51	34.73	97.41	451
20	160	147	132	98
30	110	105	88	71
40	80	72	66	49
50	65	58	48	34
60	55	47	32	20

Table 87: Functional properties of sorbitol solution.

Parameter	Observations
Refractive index (20°C)	1.45
Relative sweetness (10% solution)	60:100
Cooling effect	In 6 seconds
Thermal stability	160-180°C
Flash point	Above 300°C
Specific gravity at 25°C	1.28

Bio-ethanol production

At present, there are three important routes for producing bio ethanol:

1. From starch or sugar crops,
2. From cellulose following enzymatic hydrolysis and
3. From cellulose following chemical hydrolysis.

The cost of production of bio ethanol from the three routes is comparable. However, commercial production is based on starch and sugar crops only; there is no full-scale production plant based on cellulose.

Billions of liters of bio ethanol is produced annually mostly in Brazil (from sugarcane) and USA (from maize). The process based on sugarcane has low productivity and the enormous amount of cellulosic bagasse is either burnt or ploughed back into soil; this biomass is a potential source of bio ethanol.

Sugarcane, Sugar beet, Sweet sorghum etc. provide a good substrate. Juices from these crops contain sugar, e.g., sugarcane juice has about 12% fermentable sugars, and can be used directly for fermentation. Molasses, sweet sorghum syrup and spent sulphite liquor are the most common substrates. Molasses (C grade) obtained after sugar recovery contains about 35% sucrose, 19% other reducing agents and 14% other organic substances; thus the total fermentable sugars are nearly 50-55%. Molasses is first suitably diluted, then treated with 0.5% by weight concentrated sulphuric acid at 70-95°C (mainly for removal of calcium salts and acidification) and used for fermentation. Cereals like maize, wheat, sorghum etc. contain 60-75% w/w starch, which on hydrolysis produces glucose in the ratio 9:10 **(Figure 22)**.

Starch
(e.g. powdered seed meal, tubers etc.)

↓ Enzyme hydrolysis (50-60 °C)

Glucose

↓ Yeast Glycolysis (30-35 °C)

Pyruvate

↓ Yeast (30-35 °C) (Pyruvate decarboxylase)

Acetaldehyde

↓ Yeast (30-35 °C) Alcohol dehydrogenase

Ethanol
(Recovered by distillation)

Figure 22: Flow chart for the production of ethanol from a starch substrate.

Generally, starch is a mixture of amylose (20-30%; water soluble linear polymer) and amylopectin (70-80%; water insoluble branched polymer). A cooking process aided by enzymes usually achieves conversion of starch into glucose. Seeds are powdered and the resulting meal is mixed with water in the ratio of 1:2.5 - 3. α-amylase is then added and the temperature is raised by steam injection to 105-110ºC and held for about 20 min. During this period, starch grains become dissolved to form a viscous suspension (gelatinization) due to the cooking effect of heat. At the same time α-amylase digests partially some of the starch molecules; this reduces the viscosity of the suspension; this is called liquefaction. The suspension contains solubilized starch molecules, and dextrins (short chains of glucose units). The suspension called mash is cooled to 85-90ºC and more α-amylase is added to complete the process of liquification (for 90 min). The total dose of α-amylase is about 1.5 kg/ton starch. The mash is cooled to 55-60ºC and 1.5-3.5 liter glucoamylase/ton of starch is added to produce glucose from dextrins and maltose. The reaction duration is usually 2 days, and it achieves the dextrose equivalent of 99. Saccharification by glucoamylase continues during the fermentation.

Proper fermentation of 100g glucose by selected strains of *Saccharomyces cerevisiae* and *Saccharomyces carlsbergensis* yields 45-49g bio ethanol, the theoretical limit being 51.1g. ($C_6H_{12}O_6 \rightarrow 2C_2H_5OH + 2CO_2$). For the fermentation process batch or continuous processes are used. Batch process is the most commonly used. Ammonium sulphate or Urea (N source) and a salt of phosphoric acid (P source) are added to the substrate. Most commonly the temperature during fermentation is 32-38ºC and pH is between 4.5-5.0.

As the ethanol level rises, both ethanol production and yeast cell growth are progressively inhibited, and eventually cells begin to die.

In general, yeast growth stops at 6-9% w/v ethanol, but ethanol production, at least in some strains, may continue upto 15% ethanol or higher. Ethanol inhibition can be relieved by continuous removal of ethanol. Ethanol recovery is based on distillation.

a. The broth is distilled in a beer column to yield 85% v/v ethanol.

b. The next step of rectification gives 96.5% ethanol, which is then.

c. Dehydrated to 99.4% using benzene or cyclohexane if the ethanol is to be used as a fuel blend.

The wastewater effluent can be used directly on land, used for anaerobic digestion for methane generation, evaporated to dry syrup for animal feed or fuel or used as substrate for fungal biomass production.

Nutritional qualities and keeping quality

The objective of making all food items mentioned in this book is to increase the consumption of sorghum and millets. All the above products mentioned are having better nutritional quality in terms of protein, fiber minerals and vitamins as they are made from sorghum and other millets. These products also have a slow digestibility and the products other than sweets are ideal for diabetic and obese populations of the society.

The keeping quality of the various products varies depending on the ingredients present and method of preparation. The keeping quality of breakfast foods is minimum one day, however, the batter of idli and dosa can be stored for two days in refrigerator. The keeping quality of Lunch items is minimum one day. Jowar roti can be stored for two to three days and up to six days in refrigerator. Annam, Sankati, and Ambali can be stored only for one day. The keeping quality of sorghum sweets is minimum two days, but the kesari, kheer, burfie, sweet rolls in milk can be stored for only one day and if stored in refrigerator it will be two days. Rawa laddu can be stored for fifteen days. The keeping quality of sorghum snack foods is minimum ten days, but the pakoda, can be stored for three days only. Garelu (vada), and punugulu, kudumulu can be stored for two days. The remaining items chekkalu, murukulu, fryums, and namakpara can be stored for fifteen to twenty days.

Health foods of sorghum and present market status

Sorghum with its nutritionally rich parameters (dietary fibre, slow digestibility of starch and protein, minerals and lack of gluten) makes essential health foods. Colored sorghums with the presence of polyphenols, which are known to be antioxidants, are also regarded as health foods. Antioxidants are known for improving immunity and also preventing cardiovascular diseases and cancer. Colored sorghums are gaining importance as a source of antioxidants in U.S.A.

Processed foods and instant food mixes based on jowar are receiving attention in different parts of country in India. For example instant dosa mix made from jowar is being marketed in Coimbatore in Tamilnadu. Contact details are given below. In Zaheerabad, Andhra Pradesh in Deccan Society (a NGO) is popularizing millet foods through a shop namely 'Café Ethnic'. Sorghum and all kinds of millets are used to make different traditional foods and are made available here.

The biscuits based on jowar along with other cereals and soybean marketed with the label 'Nurtrient–Diet' by Alsa agro tech private Limited, Moosarambagh, Hyderabad.

The various health foods are listed below:

1. Manna Health Mix; Marketed by Saga Food Products Pvt. Ltd., Tamil Nadu; and
2. Pusti, the Health Mix : Manufactured and Marketed by Pusti, CBM Compound, Vishakhapatnam
3. Nutriment- Diet: Biscuits Manufactured by Alsa Agro Tech Private Limited, Moosarambagh, Hyderabad.
4. Instant Jowar *dosa* mix: Karthik Cottage, Kaveri Salai, Selvapuram, Coimbatore-641 026. Phone: 98653-76186.
5. Sunira Foods Private Limited

CONCLUSIONS

Sorghum grain with its rich nutritional qualities like high dietary fibre, minerals and slow digestibility of starch and protein is having great potential for incorporation in all possible Indian foods, both traditional and novel foods. Sorghum grain is also having potential to make *rawa* /semolina, flakes and noodles apart from pops. Sorghum rawa and flour are made into different food products that are having great importance in Indian breakfast, lunch and snack food, replacing the usual fine cereal rice and wheat. Like maize, sorghum flour can be processed into different kinds of extruded products through extruder machines. For the nutritional enrichment cereal flours including sorghum are blended with soybean and protein could be enriched up to 14% in the composite flour. The composite flour is ideal to make all bakery foods like cakes and biscuits. Instant food mixes are also possible to prepare from sorghum rawa and flour. Health drinks based on sorghum malt along with other millet malts with added flavours and preservatives are prepared. Recently, health foods based on sorghum are available in market in the form of biscuits and health mixes.

FUTURE RESEARCH WORK NEEDED ON SORGHUM

1. Dehulling of moulded grains and marketing.
2. Mix flours preparations for improvement in the nutritional quality.
3. Popularization as a health and nutraceutical food.
4. Food industries should come forward to use sorghum in various food products.
5. In plant training/demonstrations to women/youths.
6. More publication for popularization through different media (Radio, TV, Magazines, Scientific Journals, Books etc).
7. To accelerate research and extension for use of refined sorghum flour for high fiber bread and biscuits.
8. Develop literature on health benefits of sorghum for wide circulation.
9. Setting up of demonstration plant for production of different sorghum food products such as bread, biscuits, malted flours, starch, beer, ethanol etc. from sorghum.
10. Seminars with group of medical doctors, lions clubs, rotary clubs, NGOs to emphasize the health benefits of sorghum in elderly diets.
11. Every person should consume sorghum at least 3 times in week and acts a link between farmers and health food store/doctors.
12. To popularize health benefits of sorghum and make available processed sorghum flour to masses.

13. Promote farmers to produce organic sorghum and popularize it for health benefits among urbanites so that farmers get better income and motivation for sorghum farming.

Action plan needed for promotion of sorghum products in India

Short term (1-3 years)	Long term (3-5 years)
Awareness creation and generic promotion of sorghum products	Generic promotion of sorghum-products through print and electronic media
Training of rural/urban individuals and/ or groups at domestic level sorghum processing for food uses	Development of sorghum varieties with specific end uses (Food/Products/ Pharmaceutical)
Human resource and entrepreneurship developments in sorghum food products	Application of recent food processing technology such as extrusion cooking, high pressure membrane separation, enzyme processing, SCF extraction, microwave heating etc. to have newer and diversified sorghum-products for food and feed also
Preparation and notification of quality standards for sorghum-food products	Development of sorghum-based specialty food products
Establishment of sorghum food enterprises for manufacturing and marketing of sorghum foods	Formation of sorghum-foods manufactures Association of India to take care of large scale sorghum-food manufacturing and marketing in India and abroad

Sorghum-foods enhance human health at an affordable cost. Hence the need is to promote sorghum-based foods for all and by all means.

REFERENCES

A.O.A.C. (2000). Official Methods of Analysis 15th Edn. Association of Official Analytical Chemist. Washington, DC.

Amerine, M.A., Pangborn, R.M. and Rossler, E.B. (1965). Principles of sensory evaluation of foods. Acad. Press New York. pp. 350-376.

Anonymous, (1995). Food and Agricultural Organization, Sorghum and Millet in Human Nutrition. FAO: Food and Nutrition Series. No. 27, ISBN – 92-51-03381-1.

Anonymous, (2010). Agriculture Statistics at a glance. Directorate of Economic and Statistics, New Delhi.

AOAC (1990) Official Methods of Analysis 15th Edn. Association of Official Analytical Chemist Washington, DC

Au, P. M. and Fields, M. L. (1981). Nutritive quality of fermented sorghum. J. Food Sci. 46: 652-654.

Badi, S. M. and Hoseney, R.C. (1976). Use of sorghum and pearl millet flours in cookies. Cereal Chem. 53: 733-738.

Bankar, J. R., Chavan, J. K. and Kadam, S. S. (1986). Effect of grain conditioning on physico-chemical properties of sorghum bhakari. J. Maharashtra Agric. Univ., 11: 221-223.

Bender, A. E. (1973). Composition of infant foods, In Nutritional and Dietetic Foods, International Textbook Co. Ltd., London, U. K., p. 64.

Beta, T., Rooney, L.w. and Waniska, R.D. (1995). Malting characteristics of sorghum cultivar. Cereal Chem. 72: 533-538.

Beta T, Rooney LW, Waniska RD (1995) Malting characteristics of sorghum cultivar. Cereal Chem 72 : 533-538

Chavan, J. K. and Kadam S. S. (1989). Nutritional improvement of cereals by fermentation. CRC in Food Sci. and Nut. 28: 349-399.

Chavan, J. K. and Salunkhe, D. K. (1984). Structure of sorghum grain. In Nutritional and processing quality of sorghum. Pp. 21-31.

Chavan, J. K., Chavan, U. D. and Nagarkar, V. D. (1989). Effect of malting and fermentation on nutritional quality of sorghum. J. Maharashtra Agric. Univ., 14: 246-247.

Chavan, U.D. (1987). "Effect of Fermentation on the Nutritional Quality of Grain Sorghum (*Sorghum bicolour* L. Moench). "M. Sc. (Agric.) Thesis, Mahatma Phule Agriculture University Rahuri, Maharashtra, INDIA. pp.i-95.

Chavan, U.D. and Chavan, J.K. (1992). Utilization of groundnut flour in preparation of sorghum bhakari. J. Maharashtra Agric. Univ., 17 (20) : 346-347.

Chavan, U.D., Chavan, J.K. and Kadam, S.S. (1988). Effect of fermentation on soluble proteins and *in-vitro* protein digestibility of sorghum, green gram and sorghum + green gram blends. J. Food Sci. 53 : 1574-1575.

Chavan, U. D. and Chavan, J. K. (1991). Utilization of malted sorghum, mung bean and black gram in preparation of bhakari. J. Maharashtra Agric. Univ., 16: 141-142.

Chavan, U.D. and Patil, J.V. (2010). Grain Sorghum Processing. IBDC. Publishers, Lucknow (India). Pp.10-15.

Chavan, U.D., Patil, J.V. and Shinde, M.S. (2009). Nutritional and Roti Quality of Sorghum Genotypes. Indonesian Journal of Agricultural Science. 10(2): 80-87.

Chavan JK, Salunkhe DK (1984) Structure of sorghum grain in nutritional and processing quality of Sorghum. Qual. Plant. Pl. Foods Human Nutr 29: pp.21-31

Chevassua-Agnes, S., Favier, J. C. and Joseph, A. (1976). Traditional technology and nutritive value of sorghum beer from camaoon (in French) Cah. Nutr. Diet. 11: 89-104.

Dayakar Rao, B., Seetharama, N., Elangovan, M., Vilas, A. Tonapi and Ratnavathi, C.V. (2007) Changing Scenarioof Millets Cultivation in India. Souvenir, Farm Fest-2007, 23-25th February at Uppalam, Puducheri.

Dayakar Rao B, Singh JP (2010) Status paper on millets in India. Background paper prepared for *National Seminar on Millets*, organized by Directorate of Millets Development, Jaipur, and Directorate of Sorghum Research, Hyderabad

Deshpande, S.P., Borikar, S.T. and Ismail, S.S. (2003). Genetic studies for Improvement of Quality Characters in Rabi sorghum using land races. J. Agric. Food. Chem. 14: 566-571.

Deshpande SP, Borikar ST, Ismail S, Ambekar SS (2003) Genetic studies for Improvement of Quality Characters in Rabi sorghum using land races. ISMN. 44.2003

Desikachar, H. S. R. (1977). Processing of sorghum and millets for versatile food uses in India. In Sorghum and millets for human food. Pp. 41-45. Symp. Proc. IACC, Vienna, 1976. Tropical Products Institute, London.

Desikachar HSR (1975) Processing of maize, sorghum and millet for food uses. Journal of Scientific and Industrial Research 34(4): 231-237

Desikachar HSR (1977) Processing of sorghum and millets for versatile food uses in India. In sorghum and millets for human food.pp.41-45. Symp. Proc. IACC, Vienna, 1976. Tropical Products Institute, London

DuBois, M., Gilles, J.K., Hamilton, J.K., Robers, P.A. and Smith, F. (1956). Colorimetric method for determination of sugars and related substances. Analyt. Chem. 28: 350-356.

Eggum BO, Monowar L, Knudsen BKE, Munck L and Axtell D (1983) Nutritional quality of sorghum and sorghum foods from Sudan. Journal of Cereal Science 1: 127-137

FAO (1995) Food and Agricultural Organization, Sorghum and Millet in Human Nutrition. FAO : Food and Nutrition Series. No. 27, ISBN – 92-51-03381-1

GOI (2011) Agriculture statistics at a glance. Directorate of Economics and Statistics, New Delhi

Gopalan, C., Sastry, B.V.R. and Balsubramanyam, S.C. (2000). Nutritive Value of Indian Foods. National Institute of Nutrition. I.C.M.R., Hyderabad.

Hahn, R. R. (1969). Dry milling of grain sorghum. Cereal Sci. Today. 14: 234-237.

Hall AJ (2000) Sorghum utilization and the livelihoods of the poor in India : A review of findings and recommendations, International Crops Res. Institute for the Semi-Arid Tropics, ICRISAT, Patancheru, p. 41

Han, J.A., BeMiler, J.N., Hamaker, B. and Lim, S.T. (2003). Structural changes of debranched corn starch by aqueous heating and stirring. Cereal Chem. 80: 323-328.

Harris, H.B., Cammings, D.G. and Burns, R.E. (1970). Tannin content and digestibility of sorghum grain as influenced by bagging. Agron. J. 62: 633-635.

Hoseney, R. C., Zelezank, K. and Abdelrahman, A. (1983). Mechanism of popcorn popping. J. Cereal Sci. 1: 43-52.

Ibrahim, E., Abdel, R. and Magdi, A.O. (2010). Changes in Chemical Composition of available corbohydrates and amino acids content during soaking and germination of Saudi sorghum cultivars. J. Saudi. Soc. for Food and Nutrition. 5: 2.

Jambunathan, R. and Mertz, E.T. (1973). Relationship between tannin levels, rat growth and distribution of proteins in sorghum. J. Agric. Food Chem. 21: 692-696.

Karim, A. and Rooney, L. W. (1972). Characterization of pentosans in sorghum grain. J. Food Sci. 37: 369-371.

Karkalas, J., Ma, S., Morrison, W. R. and Pethrick, R. A. (1995). Some factors determining the thermal properties of amylase inclusion complexes with fatty acid. Carbohydr. Res. 268: 233-247.

Kazanas, N. and Fields, M. L. (1981). Nutritional improvement of sorghum by fermentation. J. Food Sci. 46: 819-821.

Kleih U, Bala Ravi S, Dayakar Rao B (2000). Industrial utilization of sorghum in India. Institute for the Semi-Arid Tropics, ICRISAT, Patancheru, pp. 38

Klopfenstin CF, Hoseney RC (1995) Nutritional properties of sorghum and millets. In: Sorghum and Millets: Chemistry and Technology. American Assoc. of Cereal Chemists, USA, pp. 125-168

Lowery, O. W., Rosebrough, N. J., Farr, A. L. and Randall, R. J. (1951). Protein measurement with the Folin phenol reagent. J. Biol. Chem. 193: 265-275.

McCready R.M., Guggolz, J. Silviera V., and Owens, H.S. (1950). Determination of starch and amylose in vegetables. Anal Chem. 22: 1156-1158.

Mc-Millian, W.W., Wilsemen, R.R., Burns, R.E., Harris, H.B. and Greene, G.L. (1972). Bird resistance in diverse germplasm of sorghum. Agron. J. 64 : 821-822.

Miaruddin, M., Chowdhary, M.F., Rahman, M.M. and Shahjahan, M. (2006). Standardization of spiced papads preparation from different pulses. Bangladesh J. Agril. Res. 31(4) : 681-687.

Miller, O.H. and Burns, E.E. (1970). Starch characteristics of selected grain sorghum as related to human food. J. Food Sci. 35: 666-668.

Moore, S. and Stein, W. H. (1948). In: Methods Enzymol. (Eds, Colowick, S. P. and Kaplan, N. D.), Academic Press, New York, 3, 468.

Mosse, J., Huet, J.C. and Baudet, J. (1988). The Amino Acid Composition of Whole Sorghum Grain in Relation to its Nitrogen Content. American Association of Cereal Chemists, Inc. 65(4): 271.

Mosse J, Huet JC, Baudet J (1988) The Amino Acid Composition of Whole Sorghum Grain in Relation to its Nitrogen Content. American Association of Cereal Chemists, Inc. Vol. 65, No. 4, 1988 271. Robertson S.K., Perez R.A. (2006) Nutritional characteristics of sorghum. World Poultry Science Journal 53 : 48-52

Murty, D. S., Patil, H. D., Prasad Rao, K. E. and House, L. R. (1982). A note on screening the Indian sorghum collection for popping quality. J. Food Sci. Technol. 19: 79-80.

Murty, D. S. and Subramanian, V. (1982). Sorghum roti. I. Traditional methods of consumption and standard procedures for evaluation. ICRISAT, 1982, Proc, Intl. Symp. On Sorghum grain quality. 28-31 October, 1981, Patancheru, A.P. India. Pp. 73-78.

Nazni, P. and Pradeepa, S. (2010). Physico-chemical analysis and organoleptic evaluation of papads prepared from jawar millet flour. International Journal of Current Research. 3 : 033-037.

Novellie, L. (1977). Beverages from sorghum and millets in "Proc. of Symp. On Sorghum and Millets for Human Food". International Association for Cereal Chemistry. Symposium. Vienna. 1976. Tropical Products Institute, London. Pp. 73-77.

Odunfa, S. A., Nordsrom, J. and Adeniran, S. A. (1994). Development of starter cultures for nutrient enrichment of ogia West African fermented cereal gruel. Report submitted to HBVC research grants program. USAID, Washington, U.S.A.

Panse, V. S. and Sukhatme, P. V. (1965). Statistical Methods for Agricultural Workers, ICAR, New Delhi, pp. 70-72.

Panse, V. S, Sukhatme, P. V. 1967. Statistical Methods for Agricultural Workers. Indian Council of Agricultural Research New Delhi pp 70-72

Park, D. and Maga, J. A. (2002). Effects of storage temperature and kernel physical condition on popping qualities of popcorn hybrids. Cereal Chem. 79: 572-575.

Parthasarathy Rao P, Basavaraj G, Ahmad W, Bhagavatula S (2010). An analysis of availability and utilization of sorghum grain in India. Journal of SAT Agricultural Research 8 : 1-8

Patil, U. S, Dalvi, U. S. and Chavan, J. K. (2003). Studies on the production of starch from mould infected discolored sorghum. J. Food Sci. Technol. 40: 115-117.

Patil, P. B., Sajjanar, G. M., Biradar, B. D. Patil, H. B. and Devarnavadagi, S. B. (2010). Technology of *hurda* production by microwave oven. Journal of Dairying, Foods and H. S. 29: 232-236.

Raghavandra Rao, S. N. and Desikachar, H. S. R. (1964). Pearling as a method of refining jower and wheat and its effect on their chemical composition. J. Food Sci. Technol. 1: 40-42.

Raghavandra Rao, S. N., Malleshi, N. G., Sreedharamurty, S., Viraktamath, C. S. and Desikachar, H. S. R. (1979). Characteristics of roti, dosa and vermicelli from maize, sorghum and bajra. J. Food Sci. Technol. 16: 21-24.

Rao Prasada, K. E. and Murty, D. S. (1981). Sorghum for Special Uses. Proceedings of the International Symposium on Sorghum Grain Quality. 129-134.

Ratnavathi CV, Bala SR, Subramanian V, Rao NS (2000) A study on the suitability of Unmalted Sorghum as a Brewing Adjunct. Journal of the Institute of Brewing 106: 383-387

Ratnavathi CV, Gopalkrishna D, Hariparasema K (2010) Compositional grain quality of sorghum hybrids evaluated under advanced hybrid Trial. Annual Report–2010. Book-2 of AICSIP Co-ordinating Team Report agn-2010. pp. 55-59

Reeve, R. M. and Walker, H. G. (1969). The microscopic structure of popped cereals. Cereal Chem. 46: 227-241.

Reichert, R. D. and Young, C. G. (1976). Dehulling cereal grains and grain legumes for developing countries. I. Quantities comparison between attrition and abrasive type mills. Cereal Chem. 53: 829-839.

Robertson S.K., Perez R.A. (2006) Nutritional characteristics of sorghum. World Poultry Science Journal 53 : 48-52

Robertson SK, Perez-Maldonado RA (2006) Nutritional characteristics of sorghum from QLD AGd NSW. Aust. Poultry Sci Symp 2006-18

Sajjanar, G, M., Patil, P. B., Biradar, B. D., Hemalatha, S., Tonapi, V. A., Elangovan, M., Raghavendra Rao K. V. and Seetharama, N. (2009). Special Sorghum Varieties of North Karnataka for Traditional and Novel Foods, Value Addition and Entrepreneurship Development, National Research Center for Sorghum,

Rajendranagar, Hyderabad and All India Co-ordinated Sorghum Improvement Project, Regional Agricultural Research Station, Bijapur, 586 101, Karnataka. Pages: 69 ISBN 81-89335-22-27.

Salunkhe DK, Kadam SS, Chavan JK (1977) Nutritional quality of protein in grain sorghum. Qual. Plant. Pl. Foods Human Nutr 27:187-205

Schrimshaw, N. S. (1974). Nutritional requirements for high protein foods, Proc. Int. Symp. Protein Foods and Concentrates, Central Food Technol. Res. Institute, Mysore, India, pp. 13-18.

Shepherd, A. D. (1974). How grain structure influences sorghum quality, presented to 5th E. African Cereals Res. Conf. Malawi.

Shobha, V., Kasturiba, B., Naik, R. K. and Yenagi, N. (2008). Nutritive Value and Quality Characteristics of Sorghum Genotypes. Karnataka Journal of Agriculture Science 20: 586-588.

Song, A. and Eckhoff, S. R. (1994). Optimum popping moisture content for popcorn kernel of different sizes. Cereal Chem. 71: 458-460.

Song, A., Eckhoff, S. R., Paulsen, M. and Litchfield, J. B. (1991). Effect of kernel size and genotype on popcorn popping volume and number of unpopped kernels. Cereal Chem. 68: 464-467.

Subramanian, V. and Jambunathan, R. (1984). Chemical composition and food quality of sorghum, pp. 32-47. In: Nutritional and Processing quality of sorghum (Eds.), D.K. Salunkhe, J.K. Chavan and S.J. Jadhav. Oxford and IBH Publ. Co., New Delhi. pp. 275.

Subramanian V, Jambunathan R (1982) Properties of sorghum grain and their relationship to roti quality. In International symposium on sorghum grain quality. ICRISAT, Patancheru, India pp. 280- 288

Subramanian V, Jambunathan R (1984) In Nutritional and Processing Quality of Sorghum (eds D K Salunkhe, J K Chavan and S J Jadhav) Oxford and IBH Publishing Company, New Delhi

Sullins, R. D. and Rooney, L. W. (1974). Microscopic evaluation of the digestibility of sorghum lines that differ in endosperm characteristics. Cereal Chem. 51: 134–142.

Sulllins, R. D., Rooney, L. W. and Rosenow, D. T. (1975). The endosperm structure of high lysine sorghum. Crop Sci. 15: 599-600.

Tian, Y., Buriak, P. and Eckhoff, S. R. (2001). Effect of hybrid and physical properties of individual popcorn kernels on expansion volume. Cereal Chem. 78: 578-582.

Unhale D. S., Sakhale, B. K., Ranveer, R. C. and Pawar, V. D. (2012). Studies on shelf life extension of sorghum *roti*. International Food Research Journal 19: 733-736.

Van Soest, P. J. (1963). Use of detergents in analysis of fibrous feed II. A rapid method for determination of fiber and lignin. J. Asso. Official Agril. Chem. 46: 829-835.

Van Soest, P. J., Vane, R. N. and Moore, L. A. (1967). Estimation of true digestibility of forages by *in-vitro* digestion of cell wall Xth International Cong. Pp. 438.

Vannalli S, Kasturiba B, Naik RK, Yenagi N (2008) Nutritive value and quality characteristics of sorghum genotypes. Karnataka J. Agriculture Science 20 : 586-588

Viraktamath CS, Raghavendra G, Desikachar HSR (1972) Varietal differences in chemical composition, physical properties and culinary qualities of some recently developed sorghum strains. Journal of Food Science and Technology 9(6): 73-76

Viraktamath CS, Desikachar HSR (1972) Processing of sorghum with special reference to its use as human food. In sorghum in seventies. Proceedings of a Symposium, (Eds. N.G.P. Rao and L.R. House) Oxford and IBM Publishing Co, New Delhi pp. 588

Wang, Y. D. and Fields, M. L. (1987). Germination of corn and sorghum in the home to improve nutritive value. J. Food Sci. 43: 1113-1115.

Watson, S. A. (1970). Wet milling process and products. In Sorghum production and utilization. J. S. Wall and Ross, W. M. Eds. The AVI Publishing Co. Inc, Westport, Connecticut USA. Pp. 602-626.

Winchester, M. D., Makokha, A. O. (2011). Mineral contents in grain sorghum. African Journal of Food Science Vol. 5(7), pp. 436 – 445

Yewale, K.V. (2013). Preparation of Bakery products using sorghum flour. M.Sc. (Agri.) Thesis submitted to Mahatma Phule Krishi Vidyapeeth, Rahuri, Maharashtra, India.

■■■■

Chapter – 4

Postharvest Management

1. POST-HARVEST LOSSES

It has been estimated that about 2.20 percent of jowar is lost at farmers' level during harvesting, threshing, winnowing, transportation and storage. Estimated post-harvest losses at producers' level are given in Table 1.

Table 1. Estimated post- harvest losses of jowar at producers' level.

Sr. No.	Operations Losses	Losses (percent to total production)
1.	Losses in transport from field to threshing floor	0.68
2.	Losses in threshing	0.65
3.	Losses in winnowing	0.32
4.	Losses in transport from Threshing floor to storage	0.21
5.	Losses in storage at farmers' level	0.34
	Total	2.20

To minimize post-harvest losses, the following measures should be followed.

1. The crop should be harvested on attaining physiological maturity.
2. The grain should be cleaned and dried to bring the moisture content below 9 per cent.
3. Use strong, and free from infestation packaging material for storage and transport.
4. Use proper scientific technique in storage.
5. Use pest control measures (fumigation) before storage.
6. Provide aeration to stored grain and stir grain bulk occasionally.
7. Seed should not be exposed to direct sunlight.

8. Seed should be inspected at fortnightly interval.
9. Use proper techniques while handling (loading & unloading), good and fast transport to avoid losses during transport.

Marketed surplus and Marketable surplus of jowar

It has been estimated that about 39.72 percent is the marketed surplus and 32.51 percent is the marketable surplus of jowar in our countries. The state-wise estimated Marketed surplus and Marketable surplus of jowar is given in Table 2.

Table 2: Estimated Marketed surplus and Marketable surplus of jowar.

Name of State **Marketed surplus**	**(Per cent to total production)** **Marketable surplus**
Andhra Pradesh	40.76
Bihar	11.43
Gujarat	22.57
Haryana	7.45
Karnataka	27.47
Madhya Pradesh	45.78
Maharashtra	42.47
Orissa	3.43
Rajasthan	24.90
Tamil Nadu	58.66
Uttar Pradesh	42.01
All India	**39.72**

Source: Marketable Surplus and Post-Harvest Losses of Jowar in India, 2002, Directorate of Marketing & Inspection, Nagpur.

2. HARVESTING CARE

The following harvesting care should be taken

1. Jowar grown for grain should be harvested when it attains physiological maturity.
2. Dry earheads promptly as moisture content influence keeping quality.
3. Avoid pest infestation during drying and threshing etc.
4. Dry the jowar grains sufficiently (below 9 percent) prior to packing and storing.

5. Jowar seeds should be dried in diffused sunlight in sun.
6. Pack the jowar in jute bags free from infestation and obnoxious smell.
7. Avoid harvesting during adverse weather conditions i.e. rains and overcast weather.
8. Harvesting should be done by adopting proper method.

3. GRADING

Grading is the process of sorting of produce according to the grades or classes. In case of jowar, the quality factors such as moisture content, foreign matter, other food grains, admixture of other varieties, damaged grains, immature grains, weevilled and shrivelled grains are considered while grading. The farmers, in order to improve the quality of produce and obtain better price, clean the jowar with sieves to remove the dust, broken grains and small size shrivelled grains etc. The buyers offer the price on the basis of visual inspection of the lot or available sample considering above mentioned quality factors.

3.1. Grade specifications

i. Specifications under AGMARK

Under the Agricultural Produce (Grading and Marking) Act 1937, the national standards for jowar are notified, considering the quality factors like a) moisture, b) foreign matter, c) other food grains, d) admixture of different varieties, e) damaged grains, f) immature grains, and g) weevilled and shrivelled grains.

Grade designation and definition of quality of *Kharif* jowar

A. General characteristics

Jowar shall-

a) be the dried mature grains of Sorghum vulgare pers raised in the kharif season;

b) be sweet, hard, clean, wholesome, uniform in size, shape, colour and in sound merchantable condition;

c) be free from added colouring matter, moulds,weevils, obnoxious substances, discolouration, poisonous seeds and all other impurities except to the extent indicated in the schedule;

d) Uric acid and aflatoxin shall not exceed 100 miligrams and 30 micrograms per kilogram respectively;

e) be free from rodent hair and excreta;

f) Comply with the restrictions in regard to pesticides/insecticides residue (Rule 65), poisonous metals (Rule 57), naturally occuring toxic substances (Rule 57-B) and other provisions prescribed under the Prevention of Food Adulteration Rules, 1955 and as amended from time to time.

Note

(i) In foreign matter, the impurities of animal origin shall not be more than 0.10 per cent by weight.

(ii) Ergot affected grains shall not exceed 0.05 per cent by weight in damaged grains.

B. Special characteristics

Grade designation						
Moisture Foreign matter	**Other Edible Grains**		**Damaged Grains**		**Immature and shrivelled Grains**	**Weevilled Grains**
Inorganic						
Grade I	12.00	0.10	Nil	0.50	1.50	1.00
Grade II	12.00	0.25	0.10	1.00	3.00	2.00
Grade III	14.00	0.50	0.25	2.00	4.50	5.00
Grade IV	14.00	0.75	0.25	3.00	5.00	8.00

C. Definitions

(1) "**Foreign matter**" means any extraneous matter than other food grains comprising of

(a) "**Inorganic matter**" includes metallic pieces, dust, sand, gravel, stones, dirt, pebbles, lumps or earth, clay, mud and animal filth etc;

(b) "**Organic matter**" consisting of husk, straws, weeds and other inedible grains etc.

(2) "**Other Edible grains**" means any edible grains (including oil seeds) other than the one which is under consideration;

(3) "**Damaged grains**" means grains that are sprouted or internally damaged as a result of heat, microbe, moisture or weather viz., ergot affected grains and karnal bunt grains;

(4) "**Immature and shrivelled grains**" means grains that are not properly developed;

(5) "Weevilled grains" means grains that are partially or wholly bored by insects injurious to grains but does not include germ eaten grains and egg spotted grains.

(6) "**Poisonous, toxic and/or harmful seeds**" means any seed which if present in quantities above permissible limit may have damaging or dangerous effect on health, organoleptic properties or technological performance such as dhatura (D.fastuosa linn and D. stramonium linn.) corn cokle (*Agrostemma githago* L. Machai Lallium remulenum linn.) Akra (*Vicia* species).

ii. Food Corporation of India Grade Standards

Uniform specification for jowar (Marketing season 2006-07)

The Jowar shall be dried and matured grains of *Sorghum vulgare*. It shall have uniform size, shape and colour. It shall be in sound merchantable condition and also conforming to PFA standards.

Jowar shall be sweet, hard, clean, wholesome and free from *Argemone mexicana*_and *Lathyrus sativus* (khesari) in any form, coloring matter, moulds, weevils, obnoxious smell, admixture of deleterious substances and all other impurities except to the extent indicated in the schedule below :

Schedule of specification

S. No.	Refractions Maximum Limits (%)	Maximum limit (%)
1.	Foreign matter *	1.0
2.	Other food grains	3.0
3.	Damaged grains	1.0
4.	Slightly damaged & discoloured grains	1.0
5.	Shrivelled & Immature grains	4.0
6	Weevilled grains	1.0
7.	Moisture content	14.0

*Not more than 0.25% by weight shall be mineral matter and not more than 0.10% by weight shall be impurities of animal origin.

1. The definition of the above refractions and method of analysis are to be followed as given in Bureau of India 'Standard "Method of Analysis for food grains" Nos. IS: 4333 (Part-I): 1996 and IS: 4333 (Part-II):2002 and "Terminology for food grains" IS : 2813-1995 as amended from time to time.
2. The method of sampling is to be followed as given in Bureau of Indian Standard `Method of sampling of Cereals and Pulses ` No. IS: 14818-2000 as amended from time to time.

3. Within the overall limit of 1.0% for "Foreign Matter" the poisonous seeds shall not exceed 0.5% of which Dhatura and Akra seeds (Vicia species) not to exceed 0.25% and 0.2% respectively.
4. Kernels with glumes will not be treated as unsound grains. During physical analysis the glumes will be removed and treated as organic foreign matter.

iii. Codex standards

Codex alimentarius commission (CAC): Codex Alimentarius Commission (CAC) implements joint FAO/WHO Food Standards Programme. The purpose of the CAC programme is to protect the health of consumers and ensure fair practices in the food trade. The CAC is a collection of internationally adopted food standards presented in a uniform manner. Sanitary and Phyto-Sanitary Agreement and Technical Barriers to Trade Agreement of World Trade Organization recognize standards framed by CAC with respect to safety and quality aspects of food items. Thus, for international trade, standards framed by CAC are recognized.

Codex standard for sorghum grains: CODEX STAN 172-1989 (Rev. 1 - 1995)

The Annex to this standard contains provisions which are not intended to be applied within the meaning of the acceptance provisions of Section 4.A (I)(b) of the General Principles of the Codex Alimentarius.

1. Scope

This Standard applies to sorghum grains as defined in Section 2, for human consumption; i.e., ready for its intended use as human food, presented in packaged form or sold loose from the package directly to the consumer. It does not apply to other products derived from sorghum grains.

2. Description

2.1 Definition of the Product

Sorghum grains are whole or decorticated grains obtained from species of *Sorghum bicolor* (L.) Moench. They may be suitably dried if necessary.

Whole Sorghum Grains

These are sorghum grains obtained as such after a complete threshing without any further treatment.

Decorticated Sorghum Grains

These are sorghum grains from which the external casings and whole or parts of the germ have been removed in an appropriate manner, using mechanical treatment.

3. Essential composition and quality factors

Quality Factors - General

1. Sorghum grains shall be safe and suitable for human consumption.
2. Sorghum grains shall be free from abnormal flavours, odours, and living insects.
3. Sorghum grains shall be free from filth (impurities of animal origin, including dead insects) in amounts which may represent a hazard to human health.

Quality Factors - Specific

1. Moisture Content 14.5% m/m max.

Lower moisture limits should be required for certain destinations in relation to the climate, duration of transport and storage. Governments accepting the Standard are requested to indicate and justify the requirements in force in their country.

2. Definition of Defects

The product shall have not more than 8.0% total defects including extraneous matter, inorganic extraneous matter, and filth as contained in the standards and blemished grains, diseased grains, broken kernels, and other grains as contained in the Annex.

Extraneous matter is all organic and inorganic material other than sorghum, broken kernels, other grains and filth. Extraneous matter includes loose sorghum seed coats. Sorghum grains shall have not more than 2.0% extraneous matter of which not more than 0.5% shall be extraneous inorganic matter.

Filth is impurities of animal origin including dead insects (0.1% m/m max).

Toxic or noxious seeds

The products covered by the provisions of this standard shall be free from the following toxic or noxious seeds in amounts which may represent a hazard to human health.

Crotolaria (*Crotalaria* spp.), Corn cockle (*Agrostemma githago* L.), Castor bean (*Ricinus communis* L.), Jimson weed (*Datura* spp.), and other seeds that are commonly recognized as harmful to health.

Tannin Content

a. For whole sorghum grains, the tannin content shall not exceed 0.5% on a dry matter basis.

b. For decorticated sorghum grains, the tannin content shall not exceed 0.3% on a dry matter basis.

4. CONTAMINANTS

1. Heavy Metals

Sorghum grains shall be free from heavy metals in amounts which may represent a hazard to human health.

2. Pesticide Residues

Sorghum grains shall comply with those maximum residue limits established by the Codex Alimentarius Commission for this commodity.

3. Mycotoxins

Sorghum grains shall comply with those maximum mycotoxin limits established by the Codex Alimentarius Commission for this commodity.

5. HYGINE

1. It is recommended that the product covered by the provisions of this standard be prepared and handled in accordance with the appropriate sections of the Recommended International Code of Practice -General Principles of Food Hygiene (CAC/RCP 1-1969, Rev. 2-1985, Codex Alimentarius Volume 1B) and other Codes of Practice recommended by the Codex Alimentarius Commission which are relevant to this product.
2. To the extent possible in good manufacturing practice, the product shall be free from objectionable matter.
3. When tested by appropriate methods of sampling and examination, the product:
   - shall be free from microorganisms in amounts which may represent a hazard to health;
   - shall be free from parasites which may represent a hazard to health; and
   - shall not contain any substance originating from microorganisms in amounts which may represent a hazard to health.

6. PACKAGING

1. Sorghum grains shall be packaged in containers which will safeguard the hygienic, nutritional, technological, and organoleptic qualities of the product.

2. The containers, including packaging material, shall be made of substances which are safe and suitable for their intended use. They should not impart any toxic substance or undesirable odour or flavour to the product.
3. When the product is packaged in sacks, these must be clean, sturdy and strongly sewn or sealed.

7. LABELLING

In addition to the requirements of the Codex General Standard for the Labelling of Prepackaged Foods (CODEX STAN 1-1985, Rev. 1-1991, Codex Alimentarius Volume 1A), the following specific provisions apply:

1. Name of the Product

The name of the product to be shown on the label shall be "sorghum grains."

2. Labeling of Non-Retail Containers

Information for non-retail containers shall either be given on the container or in accompanying documents, except that the name of the product, lot identification and the name and address of the manufacturer or packer shall appear on the container. However, lot identification and the name and address of the manufacturer or packer may be replaced by an identification mark, provided that such a mark is clearly identifiable with the accompanying documents.

8. METHODS OF ANALYSIS AND SAMPLING

See Codex Alimentarius Volume 13.

In those instances where more than one factor limit and/or method of analysis is given we strongly recommend that users specify the appropriate limit and method of analysis.

Factor/Discription	Limit	Method of analysis
Colour	Buyer	Visual Examination
White, pink, red, brown, orange, yellow, or any mixture of these colours. Abnormal colour, Grtains whole natural colour has been modified by bad weather conditions, contact with the ground, heat and excessive respiration. These grains may be dull, shrivelled, swollen,	Preference	

Factor/Discription	Limit	Method of analysis
puffed, or bloated in appearance.		
Ash Decorticated sorghum grains	MAX:1.5%on a dry matter basis	AOAC 923.03 ICCNo.104/1(1990) Method for the determination of ash in dereals and cereal products (Ashing at 900 C) (Type I method) ---or--- ISO 2171:1980 cereals, pulses and derived products
Protein (N x 6.25)	MIN: 7.0% on a drymatter basis	ICC 105/1 (1986) Method for the determination of Crude Protein in Cereals and Cereal Products for Food and for Feed using selenium copper catalyst (Type I method) -----or---- ISO 1871:1975 ISO 5986:1983 – animal feeding stuffs.
Fat	Max : 4.0% on a dry matter basis	AOAC 945: 38F; 920.39C ----or---- ISO 5986:1983- animal feeding stuffs – Determination of Diethyl Ether Extract
Crude fiber	Buyer Preference	ICC 113 Determination of Crude Fiber Value(Type I) -----or---- ISO 6541 (1981) Agricultural food products determination of crude fiber content-modified Scherer method.
Defects (Total) Blemished grains, (Grains which are insect of abnormal colour, sprouted, deseased, or otherwise materially damaged.	MAX: (Total) 8.0%	Visual Examination
Diseased grains. Grains made unsafe for human	MAX : 3.0% of which diseased	

Factor/Discription	Limit	Method of analysis
consumption due to decay, moulding, or bacterial decomposition, or other causes that may be noticed without having to cut the grains open to examine them.	grains must not exceed0.5%	
Defects		
Frost-damaged grains, Grains which are damaged by frost and may appear bleached or blistered and the seed coat may be peeling; Germs may appear dead or discolored.	MAX:5.0%	
broken kernels, Sorghum and pieces of sorghum which pass through a 1.8 mm round-hole sieve	MAX:5.0%	

9. ADULTERANTS AND TOXINS

Aflatoxins are the type of mycotoxins, which are derived from the fungi, which affect human health. Aflatoxins are produced by *Aspergillus flavus, Aspergillus ochraceus* and *Aspergillus paraciticus.* Contamination of Aflatoxins occurs at any stage from field to storage, whenever environmental conditions are conducive for fungi, which grow under conditions of relatively high moisture/humidity. Drying is the most effective for controlling Aflatoxin. Avoid mechanical damage to seed during harvesting, drying, and storage. Jowar should be dried sufficiently (14% moisture) before storage. It should be stored at safe moisture level. Use proper scientific storage method. Store jowar preferably in new gunny bags. If old bags used, clean and treat it with either malathion spray or fumigant like aluminium phosphide @ 1 tablet of 3 g per m space. Prevent insect infestation by adopting chemical treatment to avoid fungus contamination. Separate the infested grains.

10. GRADING AT PRODUCER'S LEVEL AND UNDER AGMARK

The scheme, Grading at producer's level was introduced in 1962-63 by the Directorate of Marketing & Inspection (DMI). The main objective of the scheme is to bring quality awareness among the producers and offer quality produce for sale. Under the scheme, the produce is subjected to simple tests and assigned a grade before sale. The programme is being implemented by the State Governments through the Agricultural Produce Markets. In India, till 31.3.2006, 2051 grading units were set up. During the year 2005-2006, about 103452.60 tonnes of Jowar valued at Rs. 6278.24 lakh was graded at producers' level.

Table 3: Progress of grading of jowar at producers' level and estimated value during 2005-2006.

Year (Jowar)	(Quantity in tonnes) Value	Value (Rs. In lacks)
2004-2005	103452.60	6278.24
2005-2006	27246.60	1836.44

Source: D.M.I., Faridabad (Agmark Grading Statistics 2005-2006)

Grading under Agmark

Grading under Agmark is carried out by the Directorate of Marketing & Inspection in accordance with the grade specifications notified by the Govt. of India under the provisions of Agricultural Produce (Grading & Marking) Act, 1937 and rules made there under. The Directorate of Marketing & Inspection has prescribed the grade standards for Jowar.

11. PACKAGING

Good packaging is necessary for easy handling, transportation and storage. The jowar is transported from field (threshing floor) to the market and storage godown in gunny bags. Good quality gunny bags either new or second hand with proper treatment is necessary to avoid spoilage and to protect the jowar from moisture and insect attack etc. For good packaging, the packages must possess following qualities:

1. It must protect jowar well.
2. It should be strong enough to sustain weight while handling and transportation.
3. It must be convenient to handle.
4. The size of the package should be restricted in such a way that it should be easy to lift and handle by a single person.
5. It should be attractive, clean and free from any infestation etc.
6. It should be marked with the description of the content viz. Commodity,

Methods of packing

1. The graded jowar should be packed in new, clean, sound and dry jute bags, cloth bags, polywoven bags, polyethylene, polypropylene, paper packages or in other food grade plastic / packaging materials.
2. The packages should be free from insect infection, fungus contamination, deterious substances and obnoxious smell.
3. Each package should contain jowar of one grade only.

4. Each package should securely close and suitably sealed.
5. Jowar shall be packed in quantities as specified under the provisions of the standards of Weights and Measures (Packaged Commodities) Rules, 1977 as amended from time to time.
6. Suitable number of consumer packs containing graded material of the same lot may be packed in master container.

Availability of packaging material

Jowar is packed in the bags of following material

1. Jute bags
2. H.D.P.E./P.P.Bags
3. Polythene impregnated jute bags
4. Cloth bags for seed

Jute bags vs. H.D.P.E. bags: Jute is biodegradable material, while synthetic bag is not environment friendly. The disposal of unserviceable jute bags is easy compared to synthetic bags. Summary of comparative properties of HDPE (High Density Poly Ethylene) and jute bags is given as under:

Properties	HDPE bags	Jute bags
Seam strength	Poor	Strong
Surface texture	Smooth	Rough
Operational convenience	Poor (With accident risk)	Good
Capacity utilization	Poor	Excellent
Stack stability	Poor	Excellent
Resistance to hooking	Poor	Fair
Drop test performance	Poor	Good
End use performance (w.r.t. bursting, damage, spillage, replacement)	Poor	Good
Grain preservation efficiency	Poor	Excellent

Source: Indian Institute of Packaging, Packaging India.

Qualities of good packaging material

1. It should be convenient in operations.
2. The packaging material must preserve the quality of produce.
3. It should be convenient to stack.
4. It should be able to prevent spillage during transit and storage.

5. It should be cost-effective.
6. It should be clean and attractive.
7. It should be biodegradable.
8. It should be helpful in reducing the marketing cost by reducing the handling and retailing cost.
9. Packing material should be reusable.

12. TRANSPORTATION

The following means of transportation are used at different stages of marketing.

Means of transportation used at different stages of marketing. Stage of Marketing	Transportation by	Means of Transport
1. From field to the village market or primary market.	Farmer	By Head load, Pack animal, Bullock cart or Tractor's trolley.
2. From primary market to secondary whole sale market and miller	Traders / Millers	By Trucks, Railway wagons.
3. From miller and wholesale markets to retailer	Millers / Retailers	By Trucks, Railway wagons, Mini trucks.
4. From retailer to consumer	Consumer By Head loads, Pack animal, Bullock / Hand cart, Rickshaw.	By Head loads, Pack animal, Bullock / Hand cart, Rickshaw.
5. For Export	Exporter/Trader	By Ship, Air Cargo

Modes of Transport used

There are different modes of transport used in jowar transportation. Road and Rail transport are normally used for internal markets. However, for export and import, mainly Sea transport is used. The most common modes of transportation are;

1. Road transportation: Road transport is the most popular mode for movement of jowar to the assembling markets as well as to the distribution centers. The following means of road transport are used in different parts of the country to transport jowar:

a. Bullock carts/camel carts

Benefits

1. Suitable for small quantity of produce.
2. Cheap and easily available.
3. Easy transport for short distance.
4. Easily manufactured by village artisan.
5. Easily operated on *kaccha* road, muddy or sandy path.

b. Tractor trolley

Benefits

1. Carry larger quantity of produce than bullock carts in less time.
2. Suitable to transport produce in primary assembling markets in the absence of proper *pucca* road connecting the villages and market.
3. Farmers use their tractors as multipurpose for farm operations.

c. Trucks

The truck is the most convenient means of transport throughout the country for longer distances for bulk quantity.

Benefits

1. Suitable for long distance.
2. Comparatively easily available.
3. Quick movement.
4. Convenient during loading and unloading.
5. Provide door to door delivery.
6. Safe transport.

2. Railways: Railway is one of the most important means of transportation.

Benefits

1. Suitable for carrying larger quantity of produce.
2. Suitable for long distances throughout India.
3. Comparatively cheaper and safer mode of transport.

3. Water transport: This is the oldest and cheapest mode of transport. It includes river transport, canal transport and sea transport.

Benefits

1. Suitable for carrying large quantity for export and import to other countries.
2. Comparatively cheaper mode of transport.

Selection of mode of transport

Following points may be considered while selection of mode of transportation:

1. Mode of transport should be selected as per the requirement according to quantity and distance.
2. It should be easily available at the time of transport, particularly during peak period after harvest.
3. It should be comparatively cheaper among available alternatives.
4. It should protect jowar from adverse weather conditions.
5. It should be free from pilferage.
6. It should be insured against any accident, natural calamities, etc.
7. It should ensure delivery of goods within specified time and at specified destination.

13. STORAGE

Requirements for safe and scientific storage

Following requirements should be considered for safe and scientific storage of jowar:

1. **Selection of site:** The storage structure should be located on a raised well drained site. It should be easily accessible. The site should be free from water logging, dampness, excessive heat, insects, rodents, termites etc.
2. **Selection of storage structure:** The storage structure should be selected according to the quantity of jowar to be stored and the period of storage. In godowns sufficient space should be provided between two stacks, between stacks and walls, so that proper aeration can be available.
3. **Cleaning and fumigation:** Before, storage of jowar, godown/structure should be properly cleaned and fumigated. There should be no cracks, holes or crevices in the structure.
4. **Drying and cleaning grains:** Before storage jowar grains should be properly dried and cleaned to avoid quality deterioration.
5. **Cleaning of bags:** Always use new gunny bag. In case of second hand gunny bags, it should be disinfested by boiling in one percent Malathion Solution for 3 to 4 minutes and fully dried.

6. **Separate storage of new and old stock:** To prevent contamination from the old stock to new stock, it is advised to store them separately.
7. **Use of dunnage:** Bag of jowar should be kept on wooden crates or bamboo mats along with cover of polythene sheet to avoid absorption of moisture from the floor.
8. **Proper aeration:** Proper aeration should be provided during dry and clean weather but care should be taken to avoid aeration in rainy season to protect the stock from moisture.
9. **Cleaning of vehicles:** The vehicles used for transportation of jowar should be cleaned by phenyl to avoid infestation.
10. **Regular inspection:** To maintain proper health and hygiene of stock regular inspection of stored jowar is necessary. Periodic fumigation should be carried out in case of long storage.

14. MAJOR STORAGE PESTS AND THEIR CONTROL MEASURES

Various insects and pests during storage damage jowar. Losses due to damage are both qualitative and quantitative. It also damages viability of seed. Some of the jowar pests often start infestation in the field several weeks before the crop is harvested.

Factors affecting severity of damage

The severity of infestation depends on following factors:

1. Moisture in grain at the time of storage.

2. Relative humidity in atmosphere.
3. Temperature inside the store/container.
4. Type of storage structure used.
5. Storage period.
6. Processing method adopted.
7. Sanitation.
8. Fumigation frequency etc.

Major stored grain pests of Jowar and their control measures are as under

Name of pest	Figure of pest	Damage	Control measures
1. Rice weevil *Sitophilus oryzae* (Linn.)	 	Adults and larvae both bore into grains and feed on the grain.	Two types of treatments are followed to control infestation. **A)Prophylactic treatment :** Use following insecticide to prevent infestation in godown and stock of jowar. **1.Malathion** **(50 percent EC):** Mix 1 litre in 100 litre of water. Use 3 litre prepared solution per 100 square meter area. Spray every 15 days interval.
2. Lesser grain borer *Rhizopertha dominica* (fabr.)	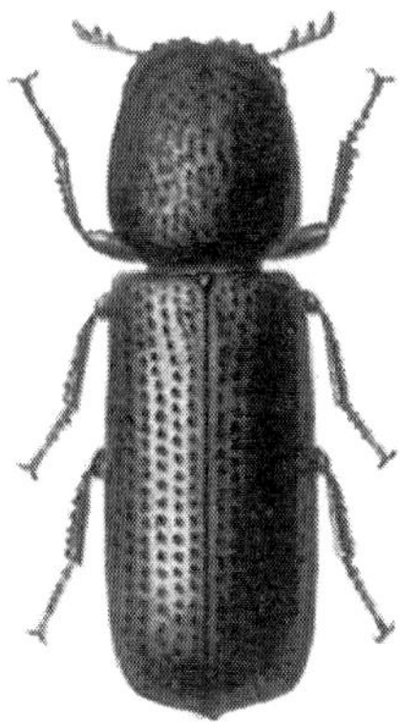	Beetles and larvae both penetrate the grain and feed. Sometimes, larvae feed on the waste flour produced by the adults. Heavy infestation makes the grain warm and moist, which leads to mould formation.	**2.DDVP (76 percent EC):** Mix 1 litre in 150 litre of water. Use 3 litre prepared solution per 100 square meter area. Do not spray on stock. Spray on walls and floors of the godown as and when

Name of pest	Figure of pest	Damage	Control measures
3. Khapra beetle *Trogoderma granarium*	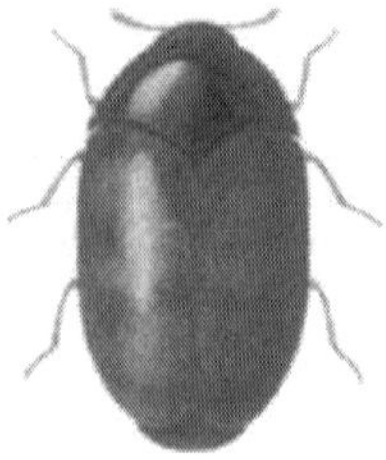 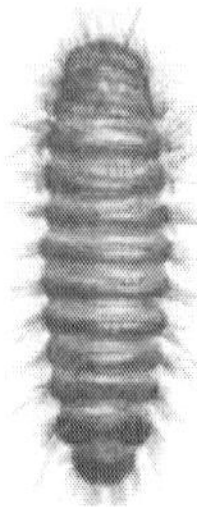	Larvae are a very serious stored pest but the beetle itself does not damage. First the larvae feed germ portion and later other parts of the grains.	required or once in a month. **3. Deltamethrin (2.5/WP):** Mix 1 kg. in 25 liter of water. Use 3 liter. prepared solution per 100 square meter area. Spray on gunny bags after 3 months' interval. **B) Curative Treatment :** Use following fumigation insecticide
4. Saw-toothed grain beetle *Oryzaephilus surinamensis* (Linn.)		Both beetle and larvae feed broken grains and damaged grains of other insects. They are usually found as a secondary pest together with other grain pests.	to control infested stock/godown of jowar in airtight condition. **1.Alluminium phosphide:** For stack fumigation use 3 tablets / tonne
5. Red rust flour beetle *Tribolium castaneum (Herbst.)*	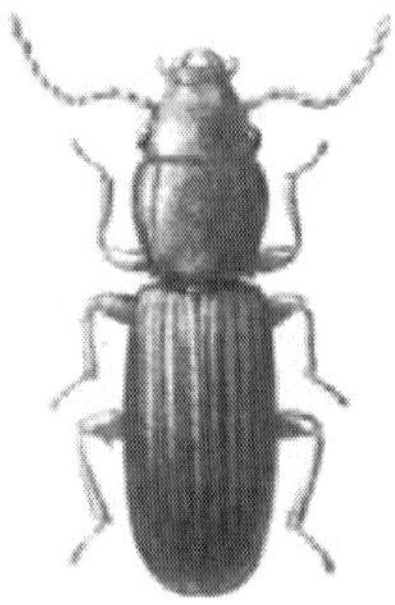	Beetle and larvae both do not cause damage to whole grain but feed on broken and damaged grains produced by milling and handling or infested /damaged grains of other insects.	and put polythene cover on infected stock. For godown fumigation, use 120 to 140 tablets per 100 cubic meter area and keep godown structure airtight and closed for 7 days. **Rat cage :** Different types of rat
6. Rice moth *Corcyra cephalonica*	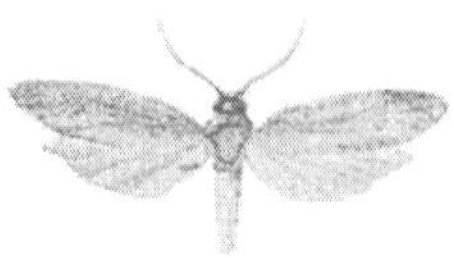	Larvae feed broken and processed jowar. Larvae produce dense webbings. Whole grain kernels are bound into lumps.	cages are available in the market. Caught rats can be killed by dipping into water. **Poison baits :** Anti-coagulant pesticide like Zinc

Name of pest	Figure of pest	Damage	Control measures
7. Rodents		Rodents eat whole grains, broken grains, flour etc. They spill more grains than they consume. Rodents also contaminate jowar by hair, urine and feces, which cause diseases like cholera, food poisoning, ringworm, rabies etc. They also damage the storage structures and other accessories of storage like wire and cable etc.	Phosphide is mixed with bread or any other food stuff used as bait. Keep baits for a week. **Rat burrow fumigation:** Put tablets of Aluminum Phosphide in each hole and burrow and block that hole by mud mixture to make it airtight.

15. STORAGE STRUCTURES

Jowar is stored to maintain the supplies between two harvests. Storage provides protection against weather, moisture, insects, micro-organisms, rats, birds and any type of infestation and contamination. In India, jowar is stored in following manner.

Traditional Storage structures	
Mud-bin	Made by bricks and mud or by straw and cow dung. These are usually cylindrical in shape with varying capacity.
Bamboo reed bin	Made by bamboo splits plastered with mixture of mud and cow dung.
Thekka	These are made up of gunny or cotton cloth wound around wooden support and generally in rectangular shape.
Metal drums	Made up with iron sheets in cylindrical and square shape with various sizes.
Gunny bags	Made up of jute.

Improved Storage Structures	
Improved bins	Different organisations developed and designed improved storage structures for scientific storage of food-grains, which are moisture resistant and rodent-proof. These are a) Pusa Kothi c) Nanda bins e)PKV bins b) PAU bins d) Hapur Kothi f) Chittore stone bins etc.
Brick-built godowns	These are made by brick-walls with cemented flooring for storing jowar in bulk and bags.
CAP (Cover and Plinth) storage	It is an economical way of storage on a large scale. The plinth is made by cement concrete and bags are stacked in open and covered by ploythene cover.
Silos	Silos are used for storage of foodgrains. These are made from concrete, bricks and metallic materials with loading and unloading equipment.

16. STORAGE FACILITIES

1. Producer's storage:

Producers store jowar in bulk at farm godown or in own house using various types of traditional and improved structures. Generally, these storage containers are used for short period. Different organisations/institutions developed improved structures for jowar storage with various capacities like Hapur kothi. Pussa bin, Nanda bin, PKV bin, etc. Other storage structures are also used for this purpose like brick-built rural godown, mud- stone godown etc. Producers also pack jowar in jute gunny bags or in gunny bags lined with polythene and stack in room.

2. Rural Godowns

Considering the importance of rural storage in marketing of agricultural produce, the Directorate of Marketing and Inspection initiated a Rural Godowns Scheme, in collaboration with NABARD and NCDC. Its objective is to construct scientific storage godowns with allied facilities in rural areas and to establish a network of rural godowns in the States and Union Territories.

The main objectives of Rural Godowns Scheme are as under:

1. To prevent distress sale of food grains and other agricultural commodities immediately after harvest.
2. To reduce quantity and quality losses arising from sub-standard storage.
3. To reduce pressure on transport system in the post-harvest period.
4. To help the farmers in getting pledge loans against the stored produce.

The statewise progress of the Rural Godown Scheme is given in Table 4.

Table 4: The state-wise progress of the Rural Godown Scheme as on 31.03.2007.

Sr. No.	State	No. of projects	Total capacity in tonnes
1.	Andhra Pradesh	735	2588759
2.	Arunachal Pradesh	1	945
3.	Assam	120	148338
4.	Bihar	292	77517
5.	Chhattisgarh	245	808297
6.	Gujarat	1887	699143
7.	Haryana	347	1517074
8.	Himachal Pradesh	31	3600
9.	Jammu & Kashmir	2	2050
10.	Karnataka	1203	981607
11.	Kerala	65	28316
12.	Madhya Pradesh	1180	1920524
13.	Maharashtra	1459	1892152
14.	Meghalaya	39	13350
15.	Nagaland	5	4700
16.	Orissa	177	375053
17.	Punjab	3483	3938789
18.	Rajasthan	445	275720
19.	Tamilnadu	330	193349
20.	Uttar Pradesh	929	2110038
21.	Uttarakhand	71	133997
22.	West Bengal	1314	459683
23.	UTs	1	4000
24.	NAFED	6	30800
25.	NCCF	1	10000
	Total	**14368**	**18217801**

Source: www.agmarknet.nic.in

3. Mandi godowns

Most of the jowar is transported to the market after the harvest. Generally, jowar is stored both in bulk and in bags. Most of the states and U.Ts have enacted Agricultural Produce Marketing Regulation Acts. The APMCs constructed storage godowns in the market yards. At the time of keeping produce in godown, a receipt is issued indicating the kind and weight of produce stored. The receipt is treated as negotiable instrument and is eligible for pledge finance. The CWC and SWCs were also allowed to construct godowns in the market yards. Co-operative societies also constructed godowns in the market yards. Both in producing and consuming centers/markets, traders also possess permanent storage in the form of godowns or warehouses, or on hire basis.

4. Central Warehousing Corporation (CWC)

CWC was established during 1957. It is the largest public warehouse operator in the country. In March 2005, CWC was operating 484 warehouses in the country. It has 16 regions, covering 225 districts, with a total storage capacity of 101.86 lakh tones. State-wise capacity with CWC as on 31.03.2005 is given Table 5.

Table 5: State-wise storage capacity with CWC as on 31-03-2005.

Name of State	No of warehouses	Total capacity (in tonnes)
Andhra Pradesh	50	1439916
Assam	6	64200
Bihar	13	97179
Chandigarh	1	13602
Chhattisgarh	10	236826
Delhi	11	181342
Goa	2	103847
Gujarat	29	622886
Haryana	25	439517
Himachal Pradesh	3	7040
Jammu & Kashmir	1	21150
Jharkhand	3	35913
Karnataka	32	453332
Kerala	9	129452
Madhya Pradesh	31	674748
Maharashtra	63	1564146
Nagaland	1	13000
Orissa	11	188206
Pondicherry	1	8940
Punjab	30	773999
Rajasthan	27	375347
Tamilnadu	26	801127
Tripura	2	24000
Uttaranchal	7	75490
Uttar Pradesh	50	1155926
West Bengal	40	685264
Total	**484**	**10186395**

Source: Central Warehousing Corporation, New Delhi.

5. State Warehousing Corporations (SWCs):

Different States have set up their own warehouses in the country. The area of operation of the State Warehousing Corporations is district places of the State. The total share capital of the State Warehousing Corporations is contributed equally by the Central Warehousing Corporation and the concerned State Government. The SWCs are under the dual control of the State Government and the CWC. As on 1st April 2005, SWCs were operating 1599 warehouses in the country with the total capacity of 195.20 lakh tonnes. The state-wise storage capacity available with SWCs as on 01.04.2005 is given below Table 6:

Table 6: State-wise storage capacity available with SWCs as on 01.04.200.

Name of SWC	Total capacity (in lakh tonnes)
Andhra Pradesh	22.82
Assam	2.48
Bihar	2.03
Chhattisgarh	6.07
Gujarat	2.27
Haryana	16.07
Karnataka	8.98
Kerala	1.92
Madhya Pradesh	11.38
Maharashtra	12.20
Meghalaya	0.11
Orissa	4.05
Punjab	60.12
Rajasthan	7.19
Tamilnadu	6.36
Uttar Pradesh	28.88
West Bengal	2.27
Total	**195.20**

Source: Central Warehousing Corporation, New Delhi.

6. Cooperatives

Cooperative storage facilities are provided to the producer at cheaper rates, which reduces the storage cost. These cooperatives also provide pledge loan against the produce and storage is more systematic and scientific than traditional storage.

Financial assistance and subsidies are provided by Government organisations/ banks to build cooperative storage.

To meet the increasing need for storage capacity, the National Cooperative Development Corporation (NCDC) encourages construction of storage facilities by cooperative, particularly at rural and market level. The number and capacity of cooperative godowns assisted by NCDC in major states are given in Table 7.

Table 7: State-wise co-operative storage facilities as on 31-3-2004.

Name of state	Rural level	Market level	Total capacity (in tonnes)
Andhra Pradesh	4003	571	690470
Assam	770	264	298900
Bihar	2455	496	557600
Gujarat	1815	401	372100
Haryana	1454	376	693960
Himachal Pradesh	1640	209	204800
Karnataka	4958	960	693590
Kerala	1959	133	323335
Madhya Pradesh	5166	1024	1305900
Maharashtra	3852	1492	2010920
Orissa	1951	595	486780
Punjab	3884	830	1986690
Rajasthan	4308	378	496120
Tamilnadu	4757	409	956578
Uttar Pradesh	9244	762	1913450
West Bengal	2834	469	483060
Other States	1046	233	644830
Total	**56096**	**9602**	**14119083**

Source: National Co-operative Development Corporation, New Delhi.

17. PLEDGE FINANCE

Micro level studies indicate that distress sale by small farmers' accounts for about 50 percent of the marketable surplus. To meet the immediate financial requirements, the farmers are often compelled to sell their produce immediately after harvest, when the prices are low. To avoid such distress sale, Government of India, promoted Pledge Finance through a network of rural godowns and negotiable warehouse receipt system. Through this scheme, small and marginal farmers can get immediate financial support to meet their requirements and retain the produce till they get remunerative price.

According to RBI guidelines, loan/advances up to 75 percent of the value of the produce stored in the godown can be advanced to farmers against pledge/hypothecation of agricultural produce including warehouse receipts subject to a ceiling of Rs. 5 lakh per borrower. Such loan shall be for a period of 6 month, which can be extended up to 12 months based on financing banks commercial judgement. The commercial/co-operative banks/RRBs provide credit to the farmers for the produce stored in the godown under this scheme. The banking institutions accept the godown receipt on its being duly endorsed and delivered to bank for pledge loan against hypothecation of produce as per RBI guidelines. Farmers are given freedom to take back their produce once the pledge loan is repaid. Facility of pledge finance is extended to all farmers, whether they are the borrowing members of Primary Agricultural Credit Societies (PACS) or not and the District Central Cooperative Banks (DCCBs) directly finance individual farmers on the strength of the pledge.

BENEFITS

1. Increase the retention capacity of the small farmers to avoid distress sale.
2. Minimises the farmers' dependence on the commission agents as the pledge finance provides financial support to them immediately after harvest period.
3. Participation of the farmers, irrespective of their land holding, helps in increasing the arrivals in market yards throughout the year.
4. Gives a sense of security to the farmers even if their produce is not sold out in market yard immediately.

REFERENCES

1. Singh, C. (1989). Modern Technologies of raising field crops.
2. National Research Centre for Sorghum (ICAR), Hyderabad publications.
3. Pandey, P.H. (1998). Principle & Practices of Post Harvest Technology.
4. Acharya, S. S. and Agarwal, N. L. (1999). Agricultural Marketing in India.
5. Pingale, S.V. (1976). Handling and Storage of Foodgrains.
6. Chakruverty A. (1988). Post Harvest Technology of Cereals, Pulses and Oil seeds.
7. Farm Machinery Research Digest, 1997, All India Co-ordinated Project on Farm Implement and Machinery, Central Institute of Agricultural Engineering, Nabi Bagh, Bhopal, India.

8. Annual Report 2003-2004, National Cooperative Development Corporation, New Delhi.

9. Annual Report, 2004-05, Central Warehouse Corporation, New Delhi.

10. Agarwal P.K. (2003). "Establishing Regional and Global Marketing Network for Smallholders' Agricultural Produce/Products with Reference to Sanitary and Phyto – sanitary (SPS) Requirement", Agricultural Marketing, April – June 2002, PP -15-23.

11. Devi Laxmi. (2003). "Inroads to contract farming" Agriculture Today, September, 2003, PP 27-35.

12. Contract farming, towards low risk and high gain agriculture, Agriculture Today, September, 2004.

13. Gururaj, H. (2002). "Contract farming: Associating for mutual benefits", www.commodityinida.com, June, 2002, pp-29-35.

14. Pandey, V. K. (2002). "Role of cooperating Marketing in India", Agricultural Marketing, Oct. – Dec. 2002 pp. 20-21.

15. Singh, H. P. (1990)."Marketing Costs, Margins and efficiency". Course material for Diploma course in Agricultural Marketing. (AMTC series – 3). Directorate of Marketing & Inspection, Branch Head Office, Nagpur.

16. Area, Production and Average Yield, 2002. Department of Agriculture and Cooperation, New Delhi.

17. Export, Import and Inter-state movement, Director General of Commercial Intelligence & Statistics (DGCIS), Kolkata.

18. Marketable Surplus and Post Harvest Losses of Jowar in India-2002. Directorate of Marketing and Inspection, Nagpur.

■■■■

Sorghum Products (Colour Photos)

Sorghum ear heads

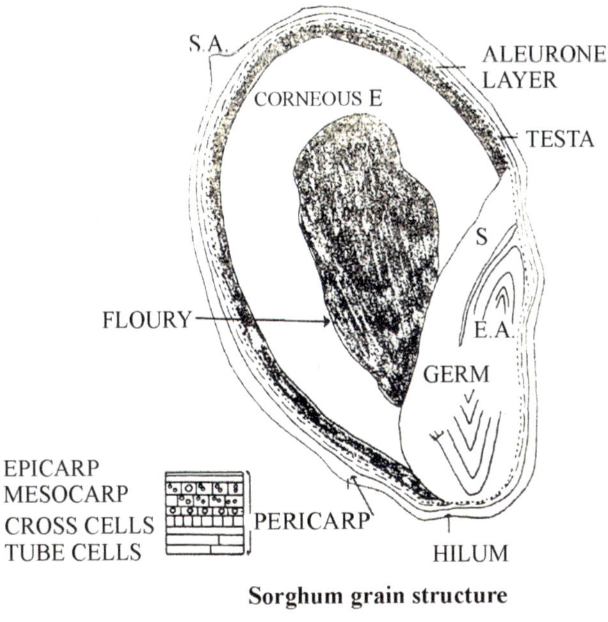

Sorghum grain structure

Flourmills

1. Vertical mill

2. Horizontal mill

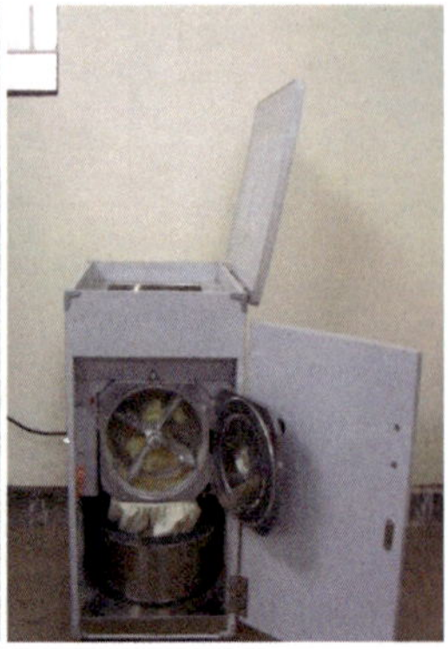
3. Pin mill

The stone mill and the pestle-mortar system used for milling cereals in India

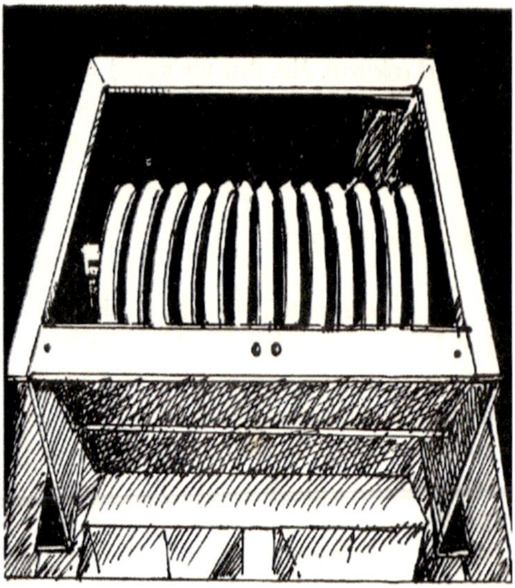
Dry milling of grain sorghum (Source: Reichert and Young, 1976)

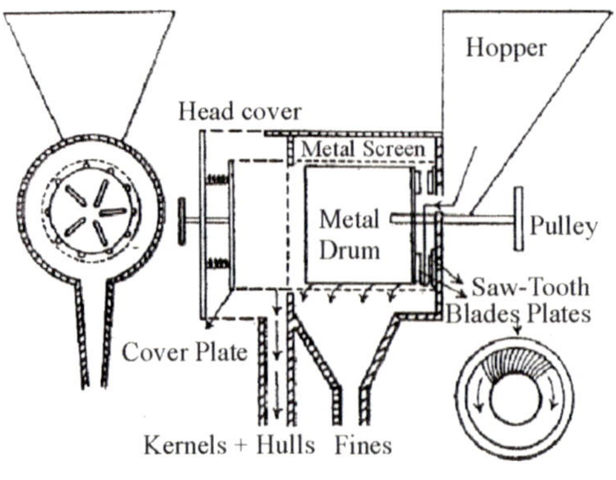

Schematic diagram of the attrition mill (Source: Reichert and Young, 1976)

Boiled sorghum grains

Sorghum pops (Phule Rohini)

Sorghum pop laddu

Sorghum *Chikki*

Sorghum flacks

Sorghum Vermicelli

Sorghum khir

Sorghum *Madak*

Sorghum *papads*

Sorghum *Upama*

Sorghum *Bhagi*

Sorghum *Shira*

Malted sorghum

Sorghum *Idali*

Sorghum masala *Dosa*

Sorghum sweet *Dosa*

Sorghum semolina *laddu*

Sorghum *Chakali*

Sorghum *Chakali*

Sorghum *Chakali*

Sorghum *Shev*

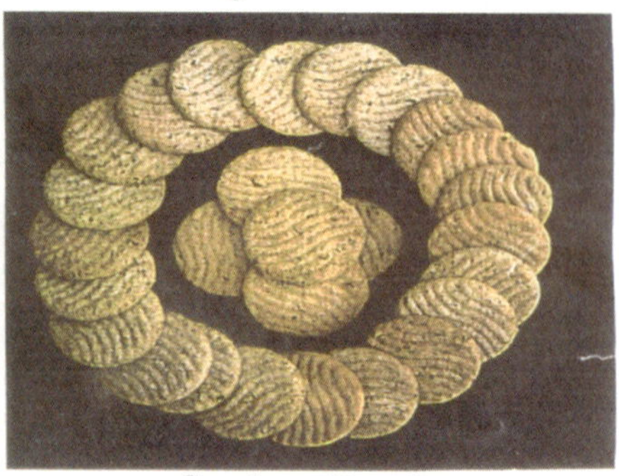

Sorghum biscuits

Sorghum *Papadi*

Sorghum *Shankarpali*

Sorghum *Cake*

Sorghum *Nankatai*

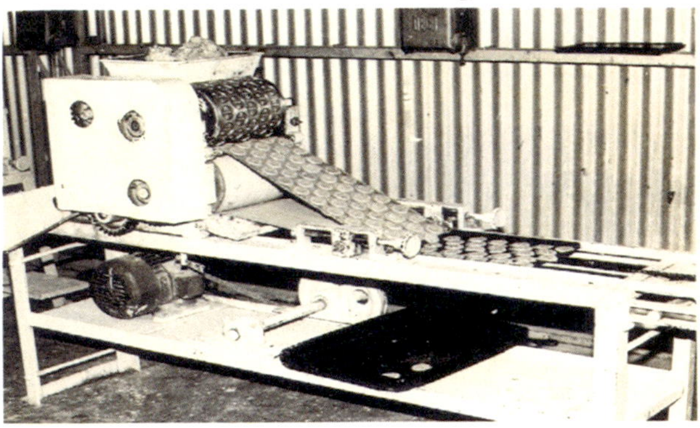

Biscuits making machine

1. Sorghum moulded grains, 2. Sound grains, 3. Starch from 1 and 4. Starch from 2

Sorghum products cafe

Appendex – I

A MODEL PROJECT REPORT FOR SORGHUM FLAKES PREPARATION

1. Land and building

A readymade shed of around 100 sq. m. May be brought which would cost around Rs. 2.00 lacks.

2. Machinery

Monthly processing capacity of 30 tonnes would require following machines:

Particulars	Quantity (Nos.)	Amount (Rs. In lacks)
Soaking tanks	4	0.40
Roaster of 250 Kg capacity	1	0.80
Cone polisher	1	0.50
Flaker	1	0.75
Sieves	3	0.75
Drier with 48 trays	1	1.00
Baby boiler of 100 kg capacity	1	0.75
Weighing scales, stitching machines, fumigation equipment set etc.	-	0.80
	Total	**5.75**

3. Miscellaneous assets

Other assets like furniture and fixtures, packing tables, plastic crates etc. would need Rs. 0.80 lacks.

4. Utilities

Daily water requirement for process and sanitation and potable purpose would be around 6000 liters. Whereas power requirement shall be 20 KW coal or LDO shall be needed for boiler.

5. Raw and packing materials

The only raw material shall be sorghum/jowar grains. Maharashtra State is a leading cultivator of sorghum followed by Madhya Pradesh, Karnataka and Andhra Pradesh and thus with proper prior arrangements, availability can be ensured. Sorghum flakes shall be packed in printed polythene bags.

6. Manpower requirements

Particulars	No	Monthly salary (Rs.)	Total monthly salary (Rs.)
Skilled workers	1	2,500/-	2,500/-
Semi-skilled worker	1	1,750/-	1,750/-
Helpers	4	1,500/-	6,000/-
Salesman	1	2,500/-	2,500/-
		Total	**13,750/-**

7. Tentative implementations schedule

Activity	Period (in months)
Application and sanction of loan	1.5
Site selection and commencement of civil work	0.5
Completion of civil work and placement of orders for machinery	1.5
Erection, installation and trial runs	0.5

8. Details of the proposed project

1. **Land and building:** A readymade shed of around 100 sq. m., would cost around Rs. 2.00 lacks as stated before.
2. **Machinery:** An estimated amount of Rs. 5.75 lacks shall be required to install required machinery.
3. **Miscellaneous assets :** Other assets would need Rs. 80,000/- as explained earlier.

4. **Preliminary and pre-operative expenses :** A provision of Rs. 1.25/- lacks would take care of pre-production expenses.

5. **Working capital requirement:** At 65% capacity utilisation in the first year, the working capital needs would be as under.

Particulars	Period	Margin	Total	Bank	Promoters
Stock of raw and packing materials	½ Month	30%	0.45	0.30	0.15
Stock of finished goods	½ Month	25%	0.60	0.45	0.15
Receivables	½ Month	25%	0.80	0.60	0.20
Working expenses	1 Month	100%	0.30	-	0.30
		Total	**2.15**	**1.35**	**0.80**

6. Cost of the project and means of financing

Items	Amount (Rs. in lacks)
Building	2.00
Plant and Machinery	5.75
Miscellaneous assets	0.80
Preliminary and pre-operative expenses	1.25
Contingencies @ 10% on building and machinery	0.80
Working capital margin	0.80
Total	**11.40**
Means of finance	
Promoter's contribution	3.20
Term loan from Bank/FI	8.20
Total	**11.40**
Debt equity ratio	2.56:1
Promoters contribution	28%

Financial assistance in the form of grant is available from the Ministry of Food Processing Industries, Govt. of India, towards expenditure on technical civil works and plant and machinery for eligible projects subject to certain terms and conditions.

9. Profitability calculations

1. **Production capacity and build-up** : As against the rated capacity of 360 tonnes, actual utilization in the first year is assumed to be 65% and thereafter it is limited to 80%.
2. **Sales revenue at 100%. :** Considering selling price of Rs. 9500/ton, the annual income at 100% activity level for 290 tonnes shall be Rs. 27.55 lacks.
3. Raw and packing materials required at 100%.

Product	Quantity (Tonnes)	Rate/Ton.	Value (Rs. in lacks)
Sorghum/Jowar	360	3750/-	13.57
Packing material	-	-	1.25
		Total	**14.82**

4. **Utilities :** The annual expenditure on utilities at 100% shall be Rs. 2.00 lacks.
5. **Interest:** Interest on term loan of Rs. 8.20 lacks is calculated @ 12% per annum assuming repayment in 5 years including a moratorium period of 1 year whereas on working capital from bank it is assumed to be 14% per annum.
6. **Depreciation:** It is calculated on work depression value (WDV) basis @ 10% on building and 15% on machinery.

10. Project profitability (Rs. in lacks)

No	Particulars	1St Year	2nd Year
A	**Installed capacity**	**360 tonnes**	
	Capacity utilization	65%	80%
	Sales realisation	17.90	22.05
B	**Cost of production**		
	Raw and packing materials	9.63	11.85
	Utilities	1.20	1.50
	Salaries	1.53	1.80
	Stores and spares	0.30	0.42
	Repairs and maintenance	0.36	0.48
	Selling expenses	0.60	0.75
	Administrative expenses	0.48	0.60

No	Particulars	1St Year	2nd Year
	Total	**14.10**	**17.40**
C	**Profit before interest & depreciation**	**3.80**	**4.65**
	Interest on term loan	0.98	0.80
	Interest on working capital	0.19	0.25
	Depreciation	1.06	0.91
	Profit before Tax	1.57	2.69
	Income Tax @ 20%	0.31	0.53
	Profit after Tax	1.26	2.16
	Cash accrual	2.32	3.07
	Repayment of term loan	-	2.05

11. Break-even point analysis

No	Particulars	Amount (Rs. in lacks)	
A	**Sales**		**22.05**
B	**Variable cost**		
	Raw and packing materials	11.85	
	Utilities (70%)	1.05	
	Salaries (70%)	1.26	
	Stores and spares	0.42	
	Selling expenses (70%)	0.53	
	Administrative expenses (50%)	0.30	
	Interest on working capital	0.25	15.66
C	**Contribution**		**6.39**
D	**Fixed cost**		**3.70**
E	**Break-even point (D ÷ C)**		**58%**

12. A. Leverages

Financial leverage:

= EBIT/EBT

= 3.74 ÷ 2.69

= 1.39

Operating leverage

= Contribution/EBT

= 6.39 ÷ 2.69

= 2.37

Degree of total leverage

= FL/OL

= 1.39 ÷ 2.37

= 0.59

B. Debt service coverage Ratio (DSCR)

Particulars	1st year	2nd year	3rd year	4th year	5th year
Cash accruals	2.32	3.07	3.30	3.58	3.84
Interest on term loan	0.98	0.80	0.55	0.30	0.18
Total (A)	3.30	3.87	3.85	3.88	4.02
Interest on term loan	0.98	0.80	0.55	0.30	0.18
Repayment of term loan	-	2.05	2.05	2.05	2.05
Total (B)	0.98	2.85	2.60	2.35	2.33
DSCR (A) (B)	3.35	1.36	1.48	1.65	1.87
Average			1.94		

C. Internal rate of return (IRR)

Cost of the project is Rs. 11.40 lacks.

Year	Cash accruals	16%	18%	20%	24%
1	2.32	2.00	1.97	1.93	1.87
2	3.07	2.28	2.20	2.13	2.00
3	3.30	2.12	2.01	1.91	1.73
4	3.58	1.98	1.85	1.73	1.51
5	3.84	1.83	1.68	1.54	1.31
6	4.29	1.76	1.59	1.44	1.18
7	4.48	1.59	1.41	1.25	0.99
	24.88	**13.56**	**12.71**	**11.93**	**10.59**

The internal rate of return is around 21%.

Appendex – II

PROJECT REPORT (STANDARD FORMAT) OF PROCESSING INDUSTRIES

1. Primary information

a. Name of Unit : ....

b. Units' address :.. ....

c. Constitution :...

d. Name of proprietor :... ....

e. Birth date, Age :.. ......

f. Qualification :... ......

g. Nature of business :.. ......

h. Registration No. of business :................................. ...

i. Loan for :... ...

j. Loan amount :... ...

2. Introduction

In this section the information on following points should be included:

a. Experience of the proprietor in this field

b. Facilities and arrangement for product sale

c. Information regarding machinery requirement and their availability as well as importance

d. Availability of the skilled and unskilled labour

e. Information regarding self-sale counter

f. Importance of this business to the society

3. Scope of business

Give the location of the processing unit and availability of raw material in that area. Provide the information regarding marketing of the processed product.

4. Manufacturing process

Give the details of manufacturing process/technology employed.

5. Land

Land for proprietor owns unit or it is in partnership give details. If in partnership then make agreement and submit with this proposal.

6. Building

Give the details about building construction (shed, RCC or load bearing etc.).

7. Water supply

Water supply from bore well or Municipal Corporation etc.

8. Electric power supply

MSEB connection deposit, main switchboard, wiring, fitting, tubes, fans, labour charges etc.

9. Printing

For labelling, advertisement, specification on the product etc.

10. Furniture

Expenditure occurred on tables, chairs, cupboards, stools etc.

11. Quantitative details (Rs. In Lacs)

Sr. No.	Description	1St year	2nd year	3rd year	4th year	5th year
1.	Installed capacity 300 days & 20 hrs					
2.	Utilization of capacity (%)					
3.	Raw material consumption at installed capacity (Rs./Ton)					
4.	Sale at installed capacity (Rate Rs. / Ton)					
5.	Raw material consumption at utilized capacity					
6.	Sales at utilized capacity					

Dehulling / Pearling Machine

Sorghum Suji

Bread, buns, cookies and biscuits prepared using sorghum flour

12. Calculation of working capital

Sr. No.	Particular	Holding period (days)	Amount
1.	Raw material		
2.	Consumable packing		
3.	Finished goods		
4.	Other Exp.		
5.	Debtors		
Gross current assets:			
6.	Less credits		
	Net working capital		
	Own margin	35 or 25%	
	Bank Finance	56 or 75%	

13. Cost of Project and Sources of Finance

Sr. No.	Item	Cost already incurred	Cost to be incurred	Total
A. Cost of project:				
1.	Land			
2.	Site development			
3.	Civil & Technical work			
4.	Plant & Machinery			
5.	Total working capital			
6.	Preliminary & preoperative Exp. (Bank fee, mortgage fee, Prof. Fee etc.)			
	Total			
B. Source of finance:				
1.	Equity capital			
2.	Subsidy/Quasi equity			
3.	Others			
4.	Unsecured Loan			
5.	Term Loan			
	Total			

14. Project Profit and Loss Account (Rs. In Lacs)

Sr. No.	Particular	1St year	2nd year	3rd year	4th year	5th year
1.	Gross sales					
2.	Job work charges					
A.	Net sales					
B. Cost of production						
1.	Raw material (as per qty. details)					
2.	Power & Fuel					
3.	Direct labour & Wages					
4.	Consumables stores					
5.	Repair & Maintenance					
6.	Other Mfg. Exp.					
7.	Depreciation					
8.	Add opening stock in process & finished goods					
9.	Deduct closing stock in process & finished goods					
C	Cost of sales					
D	Gross profit (A- C)					
E	Interest					
F	General & administrative over					
G	Profit before Tax					
H	Provision for Tax					
I	Net Profit					
J	Depression added back					
K	Net cash accruals					
L	Repayment obligations					
M	Debt service ratio (Average DSCR 3.87)					

15. Projected Balance Sheet (Rs. In Lacs)

Sr. No.	**Particular**	**1^{St} year**	**2nd year**	**3rd year**	**4th year**	**5th year**
A. Assets						
1.	Fixed assets					
2.	Investment/L. C. Deposits					
3.	Current assets					
	a. Raw material, packing Exp.					
	b. Finished goods					
	c. Debtors					
	d. Other current assets					
	Total					
B. Liabilities						
1.	Equity capital					
2.	Profit & loss A/C					
3.	Unsecured Loans					
4.	Quasi equity					
5.	Subsidy (Fr. MFPI)					
6.	Secured Loans					
7.	Working limits Fr. Bank					
1.	Term Loan					
2.	Creditors					
3.	Other current Liab. & Provision					
	Total					

16. PROJECTED CASH FLOW AND FUND FLOW STATEMENT (Rs. In Lacs)

Sr. No.	**Particular**	**1^{St} year**	**2nd year**	**3rd year**	**4th year**	**5th year**
A. Sources						
1.	Net cash accruals					
2.	Subsidy					
	Total					

Sr. No.	Particular	1^{St} year	2nd year	3rd year	4th year	5th year
B. Application						
1.	Repayment of term Loan					
2.	Repayment of unsecured & quasi equity					
3.	Increase in investment					
4.	Increase in W. C.					
	Total					

17. PROJECTED BREAKEVEN (Rs. In Lacs)

1. Fixed cost (Average)
2. Average selling price
3. Average variable cot (per ton)
   a. Materials
   b. Labours
   c. Other overheads
4. Contribution/Unit
5. Breakeven qty
6. Breakeven value
7. Percentage on installed capacity

18. REPAYMENT SCHEDULE OF TERM LOAN & INTEREST ON TERM LOAN

Period	Principle amount	Interest	Total
Interest in ….%, period in quarters ……..			
1st Year			
I Quarter			
II Quarter			
III Quarter			
IV Quarter			
Total			
2nd Year			

Period	Principle amount	Interest	Total
I Quarter			
II Quarter			
III Quarter			
IV Quarter			
Total			
3rd Year			
I Quarter			
II Quarter			
III Quarter			
IV Quarter			
Total			
4th Year			
I Quarter			
II Quarter			
III Quarter			
IV Quarter			
Total			
5th Year			
I Quarter			
II Quarter			
III Quarter			
IV Quarter			
Total			
Grand Total			

19. CALCULATION OF DEPRECIATION AS PER I. T. RULES & W.D.V. METHOD

Period	Building	Plant and Machinery	Total
Initial			
1st Year			
Total			

Period	Building	Plant and Machinery	Total
2nd Year			
Total			
3rd Year			
Total			
4th Year			
Total			
5th Year			
Total			

20. LIST OF PLANT AND MACHINERY

Sr. No.	Name of Plant or Machinery	Price (Rs.)	Quantity (Nos.)	Total amount (Rs.)
1.				
2.				
3.				
4.				
5.				
6.				
7.				
8.				
9.				
10.				
	Total			

21. TOTAL INTEREST PAYABLE YEARWISE

Period	Term	Working	Other	Total
@ p. a.	11%	15%	18%	
1st year				
2nd year				
3rd Year				
4th Year				

Period	Term	Working	Other	Total
5th Year				
Total				

22. SALARY AND WAGES DETAILS

Sr. No.	Particular	Nos.	Salary p.m.	Total
1.	Operator			
2.	Skilled workers			
3.	Unskilled workers			
	Total			

Proprietor/Partner

(Name & Address of the Unit)

■■■■

Subject Index